ATHEISM
HAS
NO
FOUNDATION

An investigative compilation of ideas and scientific - philosophical concepts, exposing the myths and dogmas of scientific atheists

Julio A. Rodríguez, IQ

Introduction

Any astute person will readily observe an abundance of anti-religious declarations propagated via the mass media in recent years. A variety of media outlets are used to "affirm" and "inform" any fabrication of facts by atheists; and much conjecture, myths, and dogma of "so called" faith (which have nothing to do with orthodox religious tradition) is included in the textbooks mandated for use in our schools.

We should like to apply the counsel attributed to the widely respected scientist **Albert Einstein** (1879 – 1955): *"if your intention is to tell the truth, do it with simplicity"*; and: *"Everything should be done as simply as possible, but not too simple."*

With this in mind I am motivated to present this modest, but concise compilation of investigated scientific facts and philosophical ideas in a manner which anyone with interest and curiosity may appreciate, though their expertise may not be in these disciplines. In doing this I have strived to remain focused on the primary purpose of all scientific analysis.

I firmly believe that in the technological era we live in and with the abundance of information easily accessed, our continual struggle to search for truth has reached a climax. As we witness a real and present psychological manipulation, this is the time to remember what the British writer and journalist **George Orwell** (1903-1950) said on one occasion: *"In an age of universal deception, speaking the truth is a revolutionary act"*

And the words of the Greek philosopher **Aristotle** (384 AC-322 AC):

"It is not enough to speak the truth; but necessary to expose the reason something is false."

My sincere hope is that those who read this book will benefit greatly. (J.R.)

3

Contents

5

8

THE RACE FOR SURVIVAL

"There are two ways to live your life:
one is as if nothing is a miracle;
the other is as if everything is a miracle"
Albert Einstein *(1879-1955)*

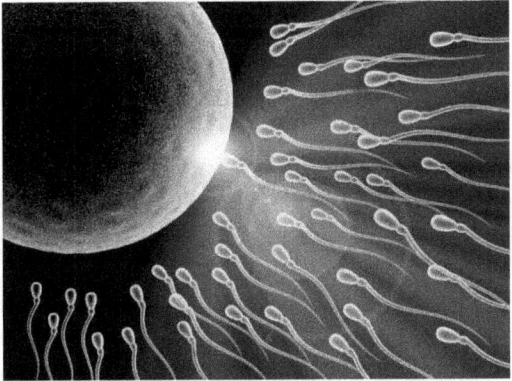

The challenge before them is unparalleled. Each competitor has an important genetic code, and each can change the course of human events.

One hundred-fifty million sperm cells will not reach their objective on time: the fertile egg; and will die trying. The one that triumphs will be the privileged: a new person.

The day has arrived. The sperm rushes through the neck of the uterus, travels the uterus and arrives at the fallopian tube. By this time most are left behind. Maybe 100 continue in the race and only one will succeed in entering and fertilizing the egg.

When this happens the ovary will become watertight, alter its chemical structure, and shut out any other sperm.

Every person that has been born on this earth is a winner, even before their birth. The sperm with its unique identity overcame millions of competitors attempting to impede its only opportunity to live.

The triumphant sperm fuses its nucleus with the ovulating egg, and the embryo forms its first cell. The fertilized embryo starts a new life journey on earth.

Genetic information transported by the victorious sperm is integrated with the genetic code of the egg and therewith we find the defining characteristics of the new person within the womb.

The environment prepared for its growth is very favorable and after the appointed time the baby will be introduced to the world: a new life!

The sperm joined the ovulating egg and each shared their genetic code and the baby developed naturally. Each respective system for its growth works to perfection. With the passing of time that person will think, reason, learn, and understand. They will be conscious of the world around them and will observe there are limitations and boundaries they must respect.

There will come a time for each person when they will ask themselves: how did all this happen? What was that race for survival? Who did they compete with? Why did they win? etc.; and realize that they were very fortunate.

On the other hand, someone may tell them that their genetic code appeared from nowhere. Their unique identity comes from nowhere and the ovulating egg, the genetic code, the joining of each code has all occurred by chance. All that was programmed and faithfully executed just happened and their growth and the world they find themselves in all came about from nowhere... what a shock!

Absurd denial of physical and biological reality

We are all aware there are two explanations of the origin of life and there are those who prefer teaching that everything existent came from nothingness.

Somehow they are satisfied and think whatever they say or do will not be held accountable to anyone; "because there is none we must give account to".

Nevertheless, for all of those persons who believe this, though they may be at peace with themselves and feel free to do as they wish, one day they will comprehend that within the genetic code they carried as they raced for survival was information programmed that would someday be their conscience.

It is the area within our soul that is profound and also where we examine our thoughts and judge whether they are right and whole, or consider the possibility of being wrong in our perceptions and how we view life.

Our mind is stirred when we discover certain scientific facts, as some of the following:

<< All living things are formed by cells, and these cells consist mostly of protein.

The probabilities that atoms and molecules unite together to form protein by chance are approximately 1 in 10^{113} (a one followed by 113 zeros). Another way of saying this is that the probability of success is one in every 10^{113} attempts;   but when a probability of success is one in every 10^{50} attempts, it is considered impossible by all mathematicians.

We must also take into account that for life to exist multiple molecules of protein are necessary. Every cell needs two thousand proteins to realize its functions which means the mathematical probability this would happen by chance is calculated to be 1 in $10^{40,000}$.

That is numerical one followed by 40,000 zeros!

The odds that cells formed themselves by chance are very remote. The probabilities they may have evolved into the array of complex life forms existent are more distantly remote. The reality is the differences between human beings and animals are so much greater than can be appreciated by the naked eye.

Unlike humans, animals have no conscience, sentiments, aesthetic sense, morality, rationale, or logic. If it is true man evolved from animals, why does this great gulf exist between humans and other species?

What should we say of the universe, of our planet earth, of the human body?

According to some astronomers the universe has 100 trillion galaxies. Each one of these galaxies may have 100 trillion stars, with most of these stars many times larger than our Sun. These galaxies do not move without order or independently, but are perfectly placed and moving in concert in an extraordinary, defined order.

Earth is a marvel that stands out among other planetary bodies. It alone is exceptional and meticulously prepared for human life, with the exact environmental conditions for human life to exist and thrive. It is an immense warehouse stocked with all the necessities: food, air, water, light and so much more.

How can anyone affirm that such a wonderful house that has been equipped and furnished for us is just by chance? With all this, can we logically insist that planet earth, having been marvelously prepared to sustain life happened merely by chance?

We will also see that within the human body 100 trillion (100.000.000.000.000) microscopic cells reside, and they are so complex as to be compared with a major city having electric generators, systems of administration, transportation, and defense. In addition, the nucleus of each cell contains DNA, the intrinsic substance that contains hundreds of thousands of genes.

It is said that a 1000 volume encyclopedia would be necessary to save all the data in the DNA of one single person. This vast amount of information would be necessary to determine the color of the person's skin, their hair, their height, and the innumerable details every individual has written genetically.

We can appreciate that every structure being built has a blueprint that is meticulously drawn up. Who designed the complex blueprint that formed the human body?

Thus far there has not been invented by man a machine, or any apparatus comparable with the organs of the human body. It is truly miraculous and the most wonderful organ we have is our brain. We read that "the central nervous system" transmits information that is more complex than the largest telephone exchange in the world.

The brain has a greater capacity than the most sophisticated computer and able to decipher and solve the most complex problem or dilemma. The difference is very pronounced" (the New Encyclopedia Britannica).

Scientists continue to marvel when researching the function of the brain. It requires much faith to believe the sufficient quantity of atoms and molecules came together precisely by chance to create this marvelous organ. >>[i], [ii]

What will we say of the axis of the earth?

<< Without the axis of the earth they would be no seasons and day and night would always be the same duration all year long. The exact amount of solar energy at each point on the Earth would be constant all year long. The earth is inclined at an angle of 23.5°.

The latitudes of the northern hemisphere receive more solar heat then the southern hemisphere sometime in June at the beginning of summer.

The days are longer and the angle of the sun greater. Meanwhile, winter begins in the southern hemisphere. The days become shorter and the angle to the Sun further. >> [iii]

With abundant reason so many renowned scientists are now expressing opposition to the assumed and celebrated conclusion that others embrace as regarding the origin of the universe. They cannot believe the absence of intelligent design and a superior power does not exist, and all that exists came **from** nothing.

The following are examples:

Sir Fred Hoyle (1915-2001), Pre-eminent mathematician, astrophysicist and author. British. He said,

"The more biochemists progress in their research and discover the profound complexity of life, it will become clearly evident that the probabilities of an accidental origin are miniscule and should be completely discarded. Life did not come into being by chance."

Charles Robert Darwin (1809 –1882), British naturalist and the "father" of evolutionary theory.

He postulated that all living species have evolved through time from a common source and by the process of natural selection.

"I have never denied the existence of God. I believe the theory of evolution is totally compatible with belief in God. The greatest argument in favor of belief in God, I perceive, is how impossible it is to demonstrate and comprehend such an immense universe, sublime beyond all measure, and mankind is the result of pure chance."

Wernher Von Braun (1912–1977), aerospace engineer, NASA astrophysicist, considered one of the most important rocket designers of the 20th century. Was the chief designer of the V – 2 rockets as well as the Saturn V, which bought man to the moon. He said:

«The more we comprehend the complexity of atomic structure, the reality of life, and the galactic universe, the more we can marvel with good reason at the splendid perfection of the divine creation».

«Over all this creation is the glory of God, who made this grand universe which man and science are continually examining and investigating day by day, with great admiration».

"The great mysteries of the universe can only validate our faith and certainty of its Creator. It is hard for me to comprehend any scientist who will not advise all, that a superior intelligence has rationally brought the universe into existence, as it is also hard to comprehend the

reasons any theologian may have for ignoring scientific advances."

Sir Isaac Newton (1642-1727), physicist, philosopher, theologian, inventor, alchemist and mathematician; father of the law of universal gravity, said:

«What we know is but a drop, but what we ignore is an immense ocean. The admirable essence of harmony in our universe would only be possible through the plan of an all-knowing and all powerful Being».

Etienne V. Borne (1907-1993), French philosopher and writer, he said:

"Atheism is the deliberate denial of the existence of God in the most definitive and dogmatic manner. It is not satisfied with relative or absolute truth but rather declares to see everything clearly and there cannot be any absolutes."

Federico Sciacca (1908-1975), Italian realism philosopher expresses his sentiments in a monologue of an atheist, saying:

<< If God doesn't exist, then why look for anything? What do I search for? I seek, and he, he whom doesn't exist, follows me persistently. He has been driven through, here, as a nail through my head. I think, and there is a nail; I think harder and it sinks deeper. My thoughts are a cruel hammer. God is merciless with the atheist. He pursues them.

Leave me God? I don't need you; remove your shadow so I may be with myself. Yours is an obstinate specter. I don't need you at all. What do you want, then, o silhouette? I denied any and all gods? No, there is no God. And after this? It is reborn as a salamander and takes the shape of a chameleon... He can be killed. I have killed him.

This specter! Specters cannot be killed. He is inside, dead, but lives. I who have killed him, have died because of him... He will not leave the dead in peace, but would resurrect them... He is alive, so alive, and feasts on the cadaver of my conscience as a vulture. He would resurrect me piece by piece.

But I would rather be dead without him then be reborn with him. Is he stronger or just foolish? My conclusion, my God is in atheism. I would not be an atheist if he did not exist.

This is the perplexing contradiction. There is no solution, but to obey. I cannot overcome, as the God that I deny confirms faith in God. My atheism must have this, as a tyrant must have their way.

Denying his existence is a forbidden hypothesis as it affirms faith. I know this, and I rebel. If you did not exist, I would not deny you. If you do exist, why is this tremendous temptation to rationalize your existence?

If you never existed, I would never have been able to think of you... Give me peace... You, who are love, are persistent as true and long-suffering love. Nothing seeks us more than love." >> [iv]

Dr. James D. Bales (1915-1995), wrote on the thoughts of the atheist:

<< An atheist believes the solid matter seen is the only reality. Matter is all that exists. This is the beginning and the end of atheistic thought. He sees the patch of agricultural land, the stars in the sky, the love of a mother, the vision of men, an insect, and the virus; and they are the manifestations of matter. They are identical in structure but different in form and organization. They are pieces of matter placed on each other and covered by other pieces of matter.

There was a time when matter existed without order and disorganized, but finally this state of disarray was converted into the orderly arrangement of our universe. It was matter in motion, without intelligent design and direction which created the present state of our universe. In addition, this lifeless matter created the living man; and thoughtless matter created the power of man to reason; as well as unconscious matter created conscious man. Matter with no moral capacity, created man and his moral sensibilities; this matter with no religion, created a man with religious aspirations.

So we now see, for anyone to embrace atheism there must be the belief in **a** Creator **that is** lifeless, has no thoughts, no conscience, no morality, no religious thoughts, and man was made without the presence of superior intelligence or purpose for the finished work we call man. Whoever believes this is extremely naïve!" >>[v]

ATHEISM

IS DEVOID OF

MORAL

FOUNDATION

ATHEISM IS DEVOID OF MORAL FOUNDATION

Atheism has led its followers into an existential vacuum. It has withheld all hope from its adherents and refused to address the basic questions of human life in a satisfactory and reasonable manner.

The failure of atheism to explain morality

<< The vision of the world presented by atheism is intellectually bankrupt and philosophically flawed. In the following article and video we will observe the inability of atheists to develop an objective rationale for morality.

First, I must say that atheists can be moral and also be persons of integrity. This is not our subject. Though we may be moral, this does not mean we have moral objectives; the behavior of an atheist can be coincidently moral in choosing to follow the mores of the culture.

This morality can be inconsistent with the concepts and behavior demonstrated by another atheist. We can conclude that there can be completely different and polar extremes of behavior where morality is subjective.

Moral objectivity is based on the external circumstances of the person. Moral subjectivity depends on each individual and their reaction within their personal situation, culture, and preferences.

21

Moral subjectivity can change and be contradictory. It can change or be contradictory and varies from person to person.

This is the best explanation offered by atheism and also how they view the world from a moral standpoint.

Meditate on the following: there is no moral concept and incorrect or correct behavior in atheism. There is no moral view of what "must be done" and "what must not be done." Why? The answer is; when God is taken away the standard of moral objectivity cannot be established. Simply stated, morality is always in a flux for the atheists.

From the atheistic viewpoint we can say that lying, deception, and stealing are not acts considered correct or incorrect. These are just spontaneous decisions that on reflection can be assigned some moralistic values if so determined by the atheist.

The atheist may conclude that we should all contribute toward a society that functions appropriately and that in general it is not beneficial for any society to practice not telling the truth, deception, or theft; but we would argue that this is a feeble attempt of rationalization, merely intellectual.

Here is the reason why. What would occur if there was a global economic catastrophe leading to such social upheaval that a common occurrence would be armed robbery to obtain basic food to live? This would then become the norm for the society in question.

Would armed robbery then be wrong? To say no is to believe a situation changes the ethics and it is acceptable. You would then not be able to complain if you are also victimized and robbed on the whim of a person in need of food, and is determined enough to point a pistol at you. We can all agree this circumstance will lead to anarchy.

Now if we do say in any circumstance an armed robbery is wrong, why is this so? If you're convinced your opinion is correct, how can you justify your opinion as an ethical principle? To assert it's wrong because you say so avoids the central issue.

It also implies there is a moral principle present that is not exclusively internal and is outside of our belief, implying a Giver of Moral Law.

In light of all this many atheists sustain the best moral system is the one that brings the most happiness, least suffering and the greatest amount of liberty for as many people as possible.

This seems a fine sentiment but unfortunately not practical or functional. We can use the institution of slavery as an example. Having a minority population in the bonds of slavery brought a great deal of happiness to a multitude of people. In this manner the greatest amount of contentment and liberty was reserved for those who owned slaves.

The atheists may submit it wrong for a majority of people to benefit from slavery, but why is it so? The reason given may be the great amount of suffering caused, but why should this suffering be wrong? This had to be oppressive and incorrect, says the atheist.

Why is it morally wrong to enslave a minority for the benefit of the majority? Atheism cannot offer us a satisfactory reply as it has no solid answers.

Let me reiterate here and remind you that atheism only offers moral subjectivity based on personal experience, circumstances and human reason. By its very nature, moral evaluation is relative, dangerous, variable, contradictory, and can easily lead to anarchy.

Although we may agree on concepts or laws that would deter anyone with a pistol from stealing for food, this is certainly not truthful morality. There is so much more and the answers are found in the Bible.

The Scriptures show us an objective moral code: do not lie, nor steal, nor commit adultery, nor give false witness, etc. This moral code does not change depending upon our opinion, circumstances, or personal preferences.

The character of God is here revealed and His unchanging Word is given. As he cannot change; so His law, which reveals universal moral concepts cannot change.

Thus, it will always be wrong to lie, steal, commit adultery, give false testimony, and this contrasts to the moral vacuum in which atheism commits itself to. Its morality is subjective.

Again, as in our prior example, if we are in an economic crisis and a strange man approaches us with a pistol in a dark street as we are carrying groceries for our family, who would you rather run into?

Would it be a believer in Jesus Christ who believes robbery is wrong and God is watching? Or would you rather confront the atheist faced with hunger, no moral compass, and willing to adapt his morality to the present circumstances of the moment? >>[vi]

Where is the atheist in the midst of suffering?

<< As we observe a suffering humanity, we can see a dilemma of cosmic proportions. It opens the door to hundreds of questions. We can relate these questions to the scientific principles of our existence and a counterpoint develops in conflict with the argument of intelligent design.

24

For example, as a scientist can assure us the eye has an excellent design but is not perfect, the philosopher can also state that we live in an orderly world that is also capricious in nature, and that contributes to great human suffering.

From the viewpoint of those who are not deists, faith in religion does not give satisfactory answers and in essence offers only a dose of superstition as bitter medicine.

They would argue that it only serves to temporarily alleviate, but does not give a concrete answer to the person who questions the reason for so much pain and suffering if an all powerful and all loving God exists there at the helm.

The problem is further exacerbated by the human experience and reality of death, the end result of pain and suffering, which according to Camus is the ultimate question of philosophy.

I find it very intriguing that with all the defamatory attacks against faith and religion, it remains the only bastion of hope as we confront our mortality, and as we confront death it is the hope of the grieving family that remains.

All of our euphemisms and abstract philosophies are not able to appease our pain and address our questions. This is where atheism is rudely challenged.

It was the perceptive viewpoint of C.S. Lewis that only human beings approach the subject of pain so uniquely. We not only confirm the reality of pain but we also place this question in a decidedly moral context, specifically the moral question of justice. Why? Why? Why?

On another occasion Lewis adamantly suggested that God may have ordained pain as his megaphone to address a morally deaf humanity.

The two related questions Lewis commented on, the reason for and purpose of pain, are always ignored by atheists as their response would puncture the core of their forceful criticism of God's existence.

In presenting the question of pain and suffering in a moral context, the atheists expose themselves as being notoriously contradictory in their perception of reality, for they simultaneously deny the existence of God.

If they cannot believe in a moral universe, how can they pose this question in a moral context? Why present the subject of pain within these parameters?

On the other hand if this is a moral universe, may it be that our suffering and pain is truly the megaphone God utilizes for humanity to be attentive and accept moral law. Now if this is a moral world, there is then possible a certain self condemnation, as the atheist is precariously trapped in the horns of a logical and moral dilemma.

If we decide it is sensible to question our condition (it is therefore also self condemning), then we may imply this is a moral universe and should not the skeptic also confront his own lack of morality? Inversely, if this is an insignificant question; as any evil is not an appropriate subject in a purely materialistic world without God, the skeptic opposes himself as he criticizes God in moral terms.

Whichever way the question is asked and by whomever; it is self-defeating.

Of course, neither is religion solely justified, though it provides relief for the hopeless. It would count for naught if it was only a psychologically induced solution.

If a religious experience is accepted without profound introspection and is only an escape from a harsh reality, we can then use a biblical analogy, the casting out of devils from the demonized only opens up the psyche for seven more demons, leaving the tormented person with the pathetic illusion of a better future. Now then, if the faith we have is based on the truth and able to withstand meticulous scrutiny, the peace and hope we seek will become a reality in this life and the next.

A diligent search for truth is necessary before we submit ourselves to the pretensions of any religion. Now, there is another question. Why isn't the same scrutiny applied to philosophies that encourage us into a life without God? Simply speaking, where is atheism on the question of human suffering? >>[vii]

CS Lewis: the conversion of a philosopher

<< C. S. Lewis led a full life and had friends, books, and students. He was born in 1898 and in 1925 taught philosophy and literature at Oxford. During his lifetime he was an eminent professor, celebrated essayist and prolific writer of texts and novels. He died in 1963.

Lewis gave this explanation for his atheism; after the early death of his mother his impression of the universe was of an expanse that was terribly cold and empty, where historically all humanity participated to some degree in a sequence of crimes, wars, disease, and suffering. "if I am asked to believe everything is the result of a benevolent spirit who is all powerful and merciful, I shall be obliged to reply the testimony points in the opposite direction."

In any case, Lewis became comfortable within atheism: "the belief in a materialistic universe had an enormous attraction for a coward as me." I had limited accountability and would not be trapped by any tragedy, as death finished everything (...)

The Christian universe confronted me with the horror of there being no door of "Escape".

The problem of pain

Atheism was the result of the pessimistic viewpoint of the world Lewis adopted: "Some years before reading Lucretius I had already felt the force of his argument, which I believe is the most convincing in favor of atheism: if there was a God that created this world, it wouldn't have been as weak and imperfect as the one we see".

In 1940, years after his conversion Lewis was asked to write, "The problem of pain." Wouldn't it have been possible for a loving and omnipotent God to destroy evil, and allow goodness and happiness to triumph in the lives of men?

Throughout these now famous pages Lewis recognizes, "it is difficult to imagine a world wherein God is continually fixing the abuses committed by men of their own free will. We would see a world where a baseball bat would turn to paper when used as a weapon against fellow man. A world would exist where sound waves would refuse to transmit the lies and insults spoken by mankind."

It would be impossible to commit bad deeds in such a world, but it assumes free will can be abolished. Now to imagine this principle and its ultimate consequences, it would be impossible to have bad thoughts, as a part of our brain where thoughts are processed would refuse to function when evil is conceived.

Any material matter within the reach of an evil person would be subject to unexpected chemical alterations. It is for these reasons that if we were to change the natural order and eliminate suffering from the world, along with free will, we would soon discover it would be necessary to completely suppress life as we now know it."

This still does not explain the reason for pain, if there is one. Neither does this explain how God continues to be good though he permits pain. Lewis utilizes this simple explanation in attempting to explain this mystery."

Pain, injustice, and ignorance are three evils with a curious difference," he says. Injustice and ignorance can be ignored, as they live within us.

Pain is different and cannot be ignored as it's an unequivocal malady that unmasks us and reveals to all that see us that something is wrong. Here is where God's voice speaks to us through our conscience and is heard clearly in our pain; using it as His megaphone to awaken us, though we would prefer to be deaf."

Lewis then elaborates that an unjust man whom life has smiled upon will never feel the necessity to correct his erroneous conduct. On the other hand, his suffering will destroy the illusion that all is always well." The megaphone of pain used by God is without a doubt terrible as his instrument. It can lead to determined and consummate rebellion against God. It may also be the last and only opportunity for a bad person to change. The veneer of appearance is removed in our pain and replaced by the banner of truth within the fortress of a rebellious soul."

Lewis did not deny that pain could be exceedingly painful."If I had found any opening to escape from its clutches, I would have dragged myself through the sewers to escape." His purpose was to reveal the simple doctrinal truth we believe to be reasonable, that it is possible for men to purify themselves through tribulation, a tenet accepted by followers of Christ.>>[viii]

29

Is Atheism a Philosophy without Hope?

A dialogue between Dr. William Lane Craig and a man named Bill:

<< – Hello Dr. Craig,

I read your article titled "Does God exist?" and in the article you write the following:

"If God doesn't exist, that we must conclude finally that there is no hope. If there is no God, that we must conclude that there will be no hope to remedy our offenses committed in our finite existence."

I simply have to disagree with this. As an atheist I do believe we can really live with tremendous hope. By this I mean that if there is no God, then when all is finished I will not have to be held accountable. I will not have to fear presenting myself before a righteous and holy God to give account for my life. A person could live life as he chooses as a result of this and not fear divine retribution. This is the hope of an atheist. Can you refute this kind of hope?

Thank you,   Bill

– Well Bill, this is truly a novel defense for atheistic hope; the hope of escaping the judgment of God. I will admit in effect, that an atheist should hope that he or she would not fall into the hands of the living God! *(Hebrews 10:31)* This does not annul what I have written. I've identified specific reasons as to why atheism is a philosophy without hope.

If God does not exist, then we should conclude that life has no hope. If there is no God, after all is done we cannot hope to be liberated of our defects as experienced during our finite existence.

For example, *there is no hope we shall be free from evil.* Although many question how God could have created the world where there is so much evil, we can observe that up to now most of the suffering in the world is due to man's inhumanity towards man. The horrors of the last two world wars in the 20th century affectively destroyed the genuine optimism of the prior century (19th). There had been much hope in man's progress.

If God does not exist, we are then trapped without hope in a world of unjustifiable suffering and unredeemable. There is no hope we shall be free of all this evil.

I say again, if there is no God *we have no hope of ever being free from the deterioration of old age, sickness, and death.* Although it may be difficult to contemplate as university students, unless you die at a young age, one day you will all be old men and old woman, each fighting a losing battle as you age and struggle against advancing deterioration, illness, and possible dementia.

Inevitably the end will come and you shall die. There is no life after the grave. For this reason we can conclude; atheism is a philosophy without hope.

You can understand that I'm speaking of the weaknesses within our finite life. I will point out two of them in particular;

I) the existence of evil; II) growing old, sickness, and death. It seems to me that atheism offers no hope when addressing these questions.

A well-known quote offered by the famous atheist philosopher Bertrand Russell laments: man is a product of forces of which had no idea concerning the purpose of their end result, and neither passion, nor heroism, nor profound thought or sentiment can preserve life after the grave.

All that has been achieved throughout the ages, all of our devotion, all inspiration, and the crystal clear light of human genius is destined for extinction after the vast death of our solar system.

The fullness of the Temple of human conquest will inevitably be entombed beneath the ashes of a universe in ruins and all these things, if they are not totally out of the realm of our contemplation, are certain to occur in an approximate time, with all this; there is no philosophical argument to the contrary that can expect to survive.

Solely within the framework of these beliefs, solely standing on the firm foundation of inexorable desperation can we move forward with assurance and construct the dwelling place of the soul.

Sartre, Camus, and many atheists have eloquently written on the desperation that is a part of atheism. For this reason atheism offers no hope.

Ironically, on the other hand Christianity not only offers the hope of freedom from afflictions and old age, sickness and death and also offers an expectant hope you would treasure: *deliverance from the hands of a just and holy God*. This was the great realization given to Martin Luther.

The same justice God that worked for his condemnation as a sinner outside of Christ; this justice was the way of salvation for him as one who was born of Christ by faith. God himself makes your debts His own, through the righteousness of Jesus Christ, when you decide to trust in Christ as Savior and Lord." There is therefore now no condemnation for those who are in Christ Jesus" (Romans 8: 1).

The believer in Jesus Christ enjoys a hope far superior to any that can be imagined by an atheist. A Christian realizes that not only has he escaped judgment, he also has assurance of salvation. You may say that Christians have given up the ability to act with impunity, as does an atheist. I give you that; but Bill, I don't choose to behave in that manner!

When you come to Christ, God changes the desires you have and you want to live a life that is impeccable and righteous. The Bible tells us the fruit of the Spirit of God fills us with love, joy, peace, patience, gentleness, goodness, faith, meekness, temperance (Galatians 5:22). Ponder this list of personal virtues. Isn't this the type of person you long to be?

A final point: You have described the hope of an atheist. *How solid is that hope?* How deep is its foundation? Most of the atheists I speak to admit that atheism cannot be proved. Actually, many insist this is true. Then how could you be sure of the veracity of your belief?

The hope of the Christian has a firm foundation, based not only on the testimony of the Holy Spirit but also on theological grounds, and the evidence of the life of Jesus Christ and his resurrection.

There exists no solid foundation that inspires hope for the atheist, even by his own admission. What happens if the "hope" you have is fundamentally flawed? What if you are wrong? >>[ix]

The failure of atheism to explain existence

Atheism is a world view that is bankrupt and wrought with philosophical dilemmas. A major problem is the inability to explain our own existence.

<< It is obvious that we do exist and even if atheists attempt to explain this away in agreement with evolutionary thought, let's put evolution aside for the moment.. We need to take a step back and ask: where did the universe come from? Let's see: is what we observe and exists caused by something that brought it into being? Since our universe does exist: what caused it into being?

My reply is there can only be two possibilities that answer the reason for the formation of the universe; the one which is impersonal, and the other possibility is the personal. These two opposing explanations encompass every possibility. So we can conclude that it's either the first or the second. There is no third option. Let us first examine the explanation given by atheists to explain the universe as the impersonal option.

If the atheist would say the universe came about of itself and existed, then we would have to call this illogical. Something that does not exist has no nature and without nature there can be no attributes, and without attributes actions cannot be concluded, as in the case of a universe that brings itself into existence. We say then this explanation doesn't work.

If the atheist would say that the universe has always existed, this would not work either as it would mean that the universe is infinite and always was. If it was ageless, why hasn't its usable energy expired as we find should happen according to the second law of thermodynamics?

In order to reach the present time as an ageless universe that would traverse infinity throughout time, you must need to overcome the impossibility of this as has been established by science. This problem also means that there could not have been prior cycles when the universe expanded and contracted. We must conclude this is not an acceptable explanation.

If the atheist claims that matter, and/or energy in some form has always existed throughout time even before the forming of the universe, but has existed in different forms, then the same problem of traverse time throughout infinity to reach the present negates that idea.

This explanation causes another problem. If the necessary conditions were always existent to bring into being the universe with the preexistent matter therein, then the effect of that transformation using that same matter and energy within the same universe should have formed it a great, undefined and immeasurable amount of time before its actual formation.

But this could not happen because it would've meant that at the present time the universe would not have usable energy. We see the problem of entropy as well as the same problem of crossing time and infinity to reach the present. This explanation is unacceptable.

Therefore the universe, composed of matter and energy cannot be infinitely ageless and timeless in its present form, or in any other form. Then how did happen? And finally, how did we arrive where we are now?

Atheists cannot help us here. Now we'll move on to give our attention to the second option; personal causation. If we would assume there to be personal influence (*which will signify a personal being formed the universe*), we have an explanation for the universe. Let me elaborate.

A rock does not suddenly stop being a rock and become an axe head unless some action occurs that impacts it. In order for material and energy to change and form something new, there has to be an external action. We should ask ourselves: what caused the matter and energy existent to form the universe?

Whatever caused this action had to exist before the universe. It follows that the universe had a beginning in time, and since matter and energy do not spontaneously change and combine to form something new, then the best explanation for the formation of the universe is that there was a decision made that put everything in motion.

In other words, a decision was made to act at a specific point in time, and this is the best explanation for the existence of the universe. As Christians we clearly acknowledge that this decision was made by the person we know and call God.

Can you see that atheists have nothing to offer relative to the important question, how did we get here?

Atheism is unable to answer important philosophical questions regarding our own existence. Any answer given is lacking and without basis, so we must conclude their response to be conjecture and ignorance.

Finally I will address one of the standard objections that atheists communicate when presented this question. Who was it that created God?

The answer is simple; no one and nothing caused God to exist. He has always existed. He is the cause and not the effect. Think about this, there cannot be causal regression to infinity. It can be compared to an infinite amount of dominoes pushing each other down.

If you travel back in time toward infinity and attempt to find the first domino that started everything, you would never find it, for it's impossible to go back toward infinity to discover it.
This conclusion is the impossibility of infinite causal regression. In addition, this means that there was never an initial cause. If there is no first cause, we would rightly conclude that there cannot be any second, third, or successive causes and none of the dominoes would fall.

But since they are falling there had to be the first causal action, which was not caused but was the initiator in a specific point of time. This is how our universe is also. There was a reason for its formation at a specific point in time. God is the initiator of this action and He decided to create the universe in the beginning and is as we read Psalm 90:2 in our Bible, *"from everlasting to everlasting, you are God."* >>^x

Perceiving Reality

It is important to understand how people arrive at contradictory conclusions when looking for answers to specific questions, though they utilize precise scientific methods and the same sources of information. Their perception of reality is the great difference. Frank Zorrilla gives us this explanation in his book, *"Knowing God through Science":*

<< The perception of reality in human beings varies in two ways. First, there is the perception every individual has relative to their experiences in life, and second there is an objective conscience that filters these experiences.

We then say two people have distinct realities depending on the way they analyze experiences, the education they have received and the various experiences they've had throughout their lives in a conscious state of mind.

Our objective conscience is a direct product of the conditioning of our mind as we form our vision of the world using the values we've learned at home, at school, and in the social context of our daily environment.

Many times we conceive our vision of the world according to the parameters learned as children, and we bear the burden of these paradigms in our mind which become who we are and how we behave.

This is the reason why excellent instruction and education is important from an early age. Generally human beings grow and develop as individuals through social conditioning or a form of collective hypnosis; accepting only a specific interpretation of reality without thoughtful analysis.

"We form our basic perception of reality in accordance with materialistic needs we perceive through our senses and for the most part fail to recognize the sublime and intangible", says Frank Zorrilla. >>[xi]

Steven R. Covey discusses this in his widely read book, *"The 7 Habits of Highly Effective People":*

<< We all have many roadmaps in our head that can be classified in 2 principal categories: map showing *the way things are*, or realities; and maps showing *the way things should be*, or values. We use these mental maps to interpret everything that we experience. We rarely question their exactitude and are little aware they even exist.

We simply *accept as fact* that the way in which we see things corresponds to the way they actually are or should be. These assumptions give rise to our attitudes and our conduct. The way we see things as being is the source of the way we choose to think and the way we comport ourselves.

Before proceeding, I should like to invite the reader to an intellectual and emotional experience. Observe the drawing on **page 41** for a few seconds.

Now look at the figure on **page 44**and carefully describe what you see.

...

Do you see a woman? How old is she? Describe her. What is she wearing? What roles do you see her in?

It is likely you would describe the woman in the second drawing as a young person of about 25 years old, very attractive, stylishly attired, a petite nose and classic look. If you were a bachelor you would probably ask her for a date. If your business was in woman's clothing, you would likely employ her as a model.

Now what if I told you that you are wrong? What would you think if I insisted this was the sketch of a woman of 60 or 70 years old, sad, with a long nose, and is absolutely unsuitable for modeling? This is the type of person you would probably help crossing the street.

Which one of us is right? Look at the sketch once again. Are you able to see the elderly woman? Look carefully if you don't see her right away. Don't you see her long, protruding nose? Her shawl?

If we were now speaking to each other face to face we may both be discussing the drawing. You would describe what you saw and I would let you know what I saw on my part. We would continue and insist until each had made his point of clearly demonstrating what we see to each other.

Since this is not the case please go to **page 44** and examine another figure. Now return to the former sketch. Can you see the old woman now? It's important that you do this before continuing to read.

I discovered this exercise many years ago at Harvard

Business School. The instructor made use of this to demonstrate with clarity and eloquence how two people can look at the same thing, disagree, and be convinced that each is right. We're not exploring logic; but entering into psychology...

... The arguments were presented and countered, and both participants were confident and convinced of their merits... Nevertheless, in the beginning only a few attempted to examine each figure from another point of reference.

After a short time of futile discussion one of the students approached the screen and pointed to a particular line of the drawing.

"This is the collar of the young woman", he said. Another said:" No, this is the mouth of the elderly woman". Slowly but surely each one began to carefully examine the specific lines that differed and finally one student, and then another, experienced the sudden realization each different image could be respectively seen.

By means of continual, tranquil, respectful and specific communication, all of those present were able to finally understand the other point of view.

When we then took our eye off the drawing and turned quickly to glance at it once again, most of us immediately saw the image we had been convinced to look for during our observation of the previous 10 seconds...

... If only 10 seconds can have such a pronounced effect on the way we see things, what do we say about the conditioning we experience all of our lives? >>[xii]

News concerning Anthony Flew, a renowned atheist who became a believer:

<< Anthony Flew, the most influential and ironclad atheist in the world, accepts the existence of God. England \ Friday 29th of May, 2009

Being considered the most influential and ironclad atheist philosopher in the world until 2004, Anthony Flew now accepts the existence of God. In his book" *There is a God: how the most notorious atheist in the world changed his outlook"*, Flew explains the reason for this change: recent scientific investigations concerning the origin of life and DNA have revealed the existence of an "intelligent creator" he asserts.

This English philosopher was one of the most vehement atheists in the world for more than five decades. He was the author of many books and also debated the best-known Christian thinkers of his day, among these the celebrated Christian apologist C.S. Lewis.

Nevertheless, at an audience he celebrated at New York University in 2004, those who attended were shocked when Flew announced at the time he accepted the existence of God and was especially impressed with the testimony given by Christianity.

In his book, which he gave the title of, *"There is a God. How the world's most notorious atheist changes his mind"* (New York: Harper one, 2007), Flew not only develops his own arguments concerning the existence of God, but he also presents strong arguments in contrast to the points of view of renowned scientists and philosophers concerning the question of God.

His investigation led him to examine, among others, the critical works of David Hume and his application to the principle of causation and the arguments of important scientists such as Richard Dawkins, Paul Davies and Stephen Hawking.

Albert Einstein was another scientist used as a reference and his critical thought concerning God. Einstein, far from being the atheist Dawkins admits to, was clearly a believer. "Intelligent design" – What caused Flew to radically change his concept of God?

He states that the principal reasons are found in recent scientific investigations concerning the origin of life; investigations that demonstrate the existence of an "intelligent creator".

Just as he had expounded in the symposium celebrated in 2004 he says his change of mind was *"mostly based on recent DNA investigations"*: I believe that it has been demonstrated that the incredible complexity of the mechanisms found in DNA necessary to begin a life, determine there had to include a participation of superior intelligence in the function and unity of such extraordinarily different elements", he asserted.

"There are an astounding number of elements of enormous complexity that participate in this process with incredible subtlety of cooperation; accomplishing the goal of working together. The great complexity I see in the mechanisms that give rise to the origin of life is what led me to conclude there had to be participation of intelligence", added Flew.

In regards to the theory of Richard Dawkins, known as the "egoistic gene" responsible for human life, Flew categorized this as *"the supreme exercise of popular mythology"*. Of course, genes cannot be egotistical or not egotistical; just as any other unconscious entity is not able to compete with another or make choices."

"I now believe that the universe was made by an Infinite intelligence and the intricate laws of the universe clearly manifest what scientists have called the Mind of God.

I believe life and reproduction originated with a divine source," he said.

"Three dimensions point us to God"/ why do I maintain this, after defending atheism for more than half a century? The simple answer is that this was the image of the world as I had seen it, which emerged from modern science. Science emphasizes the three dimensions of nature that point to a God".

"The first is the fact that nature is subject to laws. Second is the reality of life, its intelligent order and apparent purpose for its initial formation from matter. Science has not been my exclusive guide in this journey. I was helped along by my renewed study of classical philosophical arguments.

"My decision to abandon atheism was not provoked by any new phenomena or particular argument. To tell you the truth, I have re-examined every one of my ideas and thoughts to their full extent during the last two decades.

This was a consequence of my permanent assessment regarding the nature of existence. When I finally arrived at the conclusion of the existence of God, it was not because of a change in paradigm, as my paradigm remains the same.

"This is my book." - A shower of criticism rained on Flew on the part of his colleagues for his change of mind and among them was Mark Oppenheimer, who penned the article entitled, The Change of an Atheist.

Oppenheimer characterizes Flew as a senile old man who allowed himself to be exploited by Christian evangelicals for their own purposes. In addition, he accused him of signing off on a book he never wrote.

Nevertheless, Mr. Flew, who is 86 years old responded conclusively, "My name is on the book and it represents my precise opinions. I would not permit a book to be published that has my name in which I'm not one hundred percent in agreement with."

"I needed another person to transcribe the book as I'm now 84 yrs. old; he added. That was the job of Roy Varghese. The idea of anyone manipulating me due to my age is completely incorrect.

I may be old, but it's very difficult to manipulate me. This is my book, and it represents my thoughts." he concluded.
>> [xiii]

ATHEISM HAS NO SCIENTIFIC FOUNDATION

ATHEISM HAS NO SCIENTIFIC FOUNDATION

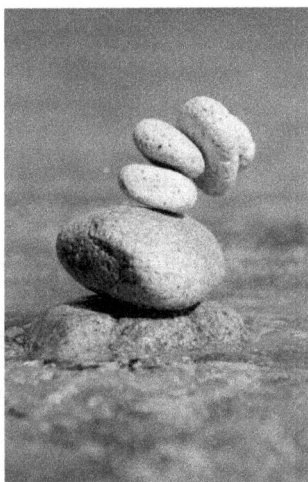

"Man finds God behind every door science is able to open" **Albert Einstein** *(1879-1955)*

"Little knowledge of science distances us from God, But much scientific knowledge returns us to Him" **Louis Pasteur** *(1822-1895)*

"Who is it, living in intimate contact with maximum possible order and divine wisdom, will not feel motivated to aspire to that which is most sublime? Who will not give adoration to the Architect of all these things? **Nicholas Copernicus** *(1473-1543)*

Anecdote about Isaac Newton:

<< How many times have we heard people affirm that faith and science are opposites? Faith vs. Reason??
But, there is no contradiction, or logical reason for us to believe in a God, since it is assumed that behind every effect there is a cause. In other words someone had to have done everything that we see surrounding us. The cause, most reasonable, is God.

Isaac Newton, one of the greatest scientists of all time, a man of reason, believed in God. Here we give a few sentences of this great man who revolutionized the science of his time:

"In the absence of other proof, the thumb alone would convince me of the existence of God"

"This beautiful system composed of the sun, planets and comets could only have been created by counsel and dominion of an intelligent and powerful being ... The Supreme God is a Being, eternal, infinite, absolutely perfect." Principia.

Once, Newton hired a clever mechanic to make a model of the solar system. Spheres representing the planets were geared together so their orbital movements reflected a realistic trajectory.

One day an atheist friend visited Newton. Seeing the model, he started it up, and exclaimed with admiration: "Who made it?" Newton said, "Nobody!"

The atheist replied: "You think I'm a fool! Of course someone has done it, and is a genius." Then Newton told his friend: "This is nothing but an insignificant imitation of a much larger system whose laws you know, and I cannot convince you that this mere toy has no designer and maker: yet you claim to believe that the great original from which the design is taken has come into existence without a designer or maker! " >>[xiv]

Errors of atheism about the origin of our universe

<< If God exists, then he is not the creator of the universe. Therefore the universe has not been created. In that case, the universe emerged from nothing, or is eternal.

The first of these alternatives is obviously absurd, because from nothing can not arise anything. We will show with an example how in their eagerness to reject the existence of God at all costs, the "wise" supporters of atheism are able to hold the most implausible claims.

In one of his many popular science books, Isaac Asimov proposed a theory about the origin of the universe from nothing, based on an analogy with the following formula: $0 = 1 + (-1)$. Just as the 0 "produces" 1 and -1, nothingness has been able to produce, in the beginning of time, a material universe and an "anti universe" (or antimatter universe).

This argument contains two gross errors:

- The ideal being "zero" is not the cause of being of ideal entities "one" and "minus one". A mathematical identity is not a causal relationship between numbers.

- There is no real correlation between the three numbers; and three entities considered real, or rather, a real entity (the universe), a hypothetical entity (the "anti-universe"); and a non-entity (nothingness). In addition, from that mathematical identity there cannot be deduced a causal relationship between these real entities.

Therefore, atheism leads to this conclusion: The universe must be eternal.

Current widespread atheist thought today is scientism or positivism. The basic premise of scientism is that the only true knowledge that man can attain derives from the particular sciences: mathematics, physics, chemistry, astronomy, geology, biology, etc. (probably the human sciences: psychology, sociology, economics, politics, history, etc. will also be included).

However, the particular sciences neither prove nor can prove that the universe is eternal, but assume it. Consequently this false assumption contradicts the fundamental principle of positivism.

This contradiction is the result of another major contradiction. The hidden point of positivism is the denial of the existence of God, although particular sciences neither prove nor can prove the nonexistence of God.

Indeed, positivism is based on (false) unscientific philosophical postulates, whose truth is assumed without any rational justification. Thus scientism, which presents itself as scientific truth, turns out to be only a false (and often subconscious) philosophy.

Contemporary science proves not only that the universe is not eternal, but even suggests very strongly the idea that the universe had an absolute beginning in time. The major consensus of current scientists supports the Big Bang theory, which implies an absolute beginning. The fact is, even if the Big Bang hypothesis is proved, science cannot prove the creation of the universe.

What happened "before" ground zero of the Big Bang, is beyond the limits of scientific knowledge, and can only be explored by theology and philosophy, which are not special sciences but universal sciences.

This means that their inquiry is based on their own methods and different from particular scientific methods, not limited to intermundane realities but aiming to achieve ultimate explanations; therefore transcendent ... >>[xv]

There are many famous scientists who have expressed their rejection of the premise of the universe coming about of nothingness; and they have confessed a belief in the Creator.

A very interesting interview

<< Project, talks about ending the battle between science and faith. Interview by Jon Sweeney

'Explorefaith' sat down recently with Dr. Francis S. Collins, M.D. Ph.D., the director of the Human Genome Project at the National Institutes of Health.

He is the author of the award-winning book, The Language of God: A Scientist Presents Evidence for Belief (Free Press; new in paperback July 2007). He is an outspoken believer in God, a Christian, and also one of the most respected scientists working today

You are Dr. Francis Collins, M.D., Ph.D., the director of the Human Genome Project—but you seem to have gained a certain notoriety as "the scientist who believes in God." Do you feel "called" to that role, at this point in history?

I am reluctant to go that far, as a claim of being "called" implies some sort of special "mission from God," and only God knows what those missions are. I have indeed been fortunate to be asked to lead a historic scientific undertaking, the Human Genome Project, and I still marvel at being chosen for this role.

One of the goals of that project has been to consider the ethical, legal, and social implications (ELSI) of these rapid advances in genetic research.

Since most Americans are believers, it has been natural to include some theological reflections in the ELSI program as well, and my own musings about science and faith could be considered part of that tradition.

Many scientists as myself believe in God, but in general we have been rather quiet about our beliefs. I do think that we are at a critical time, however, especially in the United States, in deciding how we are going to seek truth and meaning in life in the 21st century.

Clearly we will need science to help solve a lot of our problems—of illness, of communication systems, of care of our planet.

But a purely materialist approach, stripping away the spiritual aspect of humanity, will impoverish us—after all, that has been already tried (in Stalin's USSR and Mao's China) and found to be devastating. All truth is God's truth, and therefore God can hardly be threatened by scientific discoveries.

We humans have started this battle between science and faith, and it's up to us to end the battle. If I can contribute in some small way to rediscovering that harmony, then I will feel truly blessed.

You have said that DNA is "God's language." Do you mean that in a literal, or more metaphorical, sense?

A little of both. I believe that the universe was created by God with the specific intention of giving rise to intelligent life. Given that we observe DNA to be the information molecule of all living things, one can regard therefore it as the "Logos" that God has used to speak life into being...

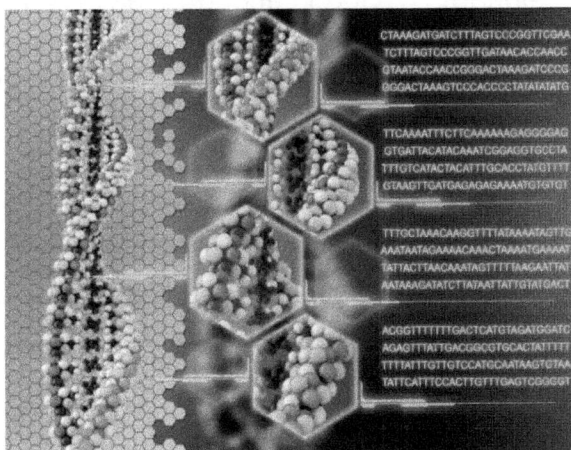

As a scientist, you test your assumptions and beliefs. But as a Christian, you have said that you took "a leap of faith." Why the two different paths?

Maybe they aren't that different. Both science and faith are ways of seeking the truth. Science seeks truth about how the natural world works, and faith seeks answers to more profound questions such as, Why is there something instead of nothing?, or What is the meaning of life?, and Is there a God?

All require a certain element of faith—you can't be a scientist unless you have faith in the fact that there is order in nature, and that nature will behave in reproducible and predictable ways.

When I was an atheist and I decided to explore the rational underpinnings of belief in God, I expected to find none—and was astounded to discover that there are strong arguments from nature and philosophy that point to God's existence. But those do not constitute a proof—apparently God intended to leave it up to us to make this decision. Perhaps such a leap of faith sounds rash to a committed materialist—but can you prove beauty? love?

In a commentary that you recently wrote for CNN.com, you mentioned the "40 percent of working scientists who claim to be believers." That number seems kind of startling to me. Is that true? Are many of them "in the closet"?

A famous survey done in 1917, and again in 1997, documented this percentage of belief amongst working scientists. Many people have been surprised by this statistic, and also surprised that the numbers haven't changed during the 20th century.

Why aren't we hearing more from scientists who believe? There is an unwritten taboo about discussing matters of faith in scientific circles, and believing scientists are sometimes also fearful that they will be seen as less intellectually rigorous by their colleagues if they admit to faith in God.

How do you nourish your spiritual life—daily, weekly?

I don't try to compartmentalize it. I try to spend time in prayer in the morning while the world is still quiet. But I also try to keep my spiritual side awake and alert during the day. I keep a Bible in my desk at work. To be honest, however, I am far from a role model here.

I often find at the end of the day that the inevitable urgencies have crowded out my intentions to be more balanced. And I am not currently a regular churchgoer. So it's fair to say I am still working on deepening my relationship with God, and that is a lifelong task.

Turning again to that commentary you wrote for CNN, I love your concluding sentence: "By investigating God's majestic and awesome creation, science can actually be a means of worship." I guess that means that your scientific work, itself, nourishes your spiritual life?

Absolutely. As a scientist who is also a believer, I find exploring nature also to be a way of getting a glimpse of God's mind. You can find God in the laboratory, just as much as in the cathedral.

You are a scientist who clearly loves the mysteries (that's a word I've seen you use a lot) of the physical world. Wouldn't many of your colleagues in the scientific community say that the purpose of science is to eliminate mystery as much as possible?

Of course! But there are always more to explore. And in my experience, unraveling the mystery of nature adds to one's sense of awe, rather than subtracting from it. Faith is also a way of trying to understand profound mysteries that science can't resolve—such as the meaning of life ... >>[xvi]

On another occasion, Dr. Francis Collins, said:

<< "I am a scientist and a believer and I see no conflict between the two worldviews. As director of the Human Genome Project, I have led an international consortium of scientists to decipher the 3.1 billion letters of the human genome, our DNA instruction book.

59

As a believer, I consider DNA as God's language, and the elegance and complexity of our bodies and the rest of nature, as a reflection of his plan.

During my time as a student of Chemistry, in the seventies, I was an atheist, because there was no reason to postulate the existence of truths outside of mathematics, physics and chemistry. But then I came to medical school and I found the themes of life and death beside the beds of my patients.

Challenged by one of them that asked me "What do you believe?, Doctor", I started looking for answers.

I had to admit that the science I loved so much, was powerless to answer questions such as what is the meaning of life?, Why am I here?, why mathematics are met anywhere?, if the universe had a beginning, who created it?, why the physical constants of the universe are so finely tuned and adjusted to allow for the possibility of complex life forms?, why men have a moral sense?, what happens after death?.

I had always assumed that faith was based on purely emotional and irrational arguments, and was astounded to discover, initially in the writings of the Oxford professor CS Lewis, and later in many other sources, that you could build a solid building that would support the plausibility of the existence of God, on purely rational grounds.

But reason only, cannot prove the existence of God. Faith is reason plus revelation, and the revealed part requires one to think with the spirit and mind. You have to hear the music, not just read the notes in the score.

Some have asked me if my brain has not exploded, if I can keep trying to understand how life works, using the tools of genetics and molecular biology and; at the same time, worship a creator God; if evolution and faith in God are not inconsistent, and if you can as a scientist believe in miracles like the resurrection.

Actually, I see no conflict in these questions and apparently neither do the 40 percent of scientists who profess to be believers...

...I have seen that there is a wonderful harmony in the complementary truths of science and faith. The God of the Bible is also the God of the genome. It can be found in a cathedral or in the laboratory. Investigating the majesty of God and the awesome creation, science can actually have a reason to worship... >> [xvii]

A very important research-study

Stephen C. Meyer published a very ample research with the title: " DNA and the   Origin of Life: Information, Specification, and Explanation"; which I recommend to be analyzed with diligence, to everyone who has studied Biology and/or Sciences. It is a deep analysis that clearly summarizes the scientific ambiguity on the origin of life, knowing the scopes of DNA.

Part of the material presented in that study, is the following:

<< Theories about the origin of life necessarily presuppose knowledge of the attributes of living cells. As historian of biology Harmke Kamminga has observed, "At the heart of the   problem of the   origin   of life lies a fundamental question: What is it exactly that we are trying to explain the   origin of?" Or as the   pioneering   chemical evolutionary   theorist Alexander Oparin put it, "The problem of the nature of life and the problem of its origin have become inseparable."

61

Origin-of-life researchers want to explain the origin of the first and presumably simplest—or, at least, minimally complex—living cell. As a result, developments in fields that explain the nature of unicellular life have historically defined the questions that origin-of-life scenarios must answer.

Since the late 1950s and 1960s, origin-of-life researchers have increasingly recognized the complex and specific nature of unicellular life and the biomacromolecules on which such systems depend.

Further, molecular biologists and origin-of-life researchers have characterized this complexity and specificity in informational terms.

Molecular biologists routinely refer to DNA, RNA, and proteins as carriers or repositories of "information."

Many origin-of-life researchers now regard the origin of the information in these 223 biomacromolecules as the central question facing their research. As Bernd Olaf Kuppers has stated, "The problem of the origin of life is clearly basically equivalent to the problem of the origin of biological information."

This essay will evaluate competing explanations for the origin of the information necessary to build the first living cell. To do so will require determining what biologists have meant by the term information as it has been applied to biomacromolecules. As many have noted, "information" can denote several theoretically distinct concepts.

This essay will attempt to eliminate this ambiguity and to determine precisely what type of information origin-of-life researchers must explain "the origin of." What follows will first seek to characterize the information in DNA, RNA, and proteins as an explanandum (a fact in need of explanation) and, second, to evaluate the efficacy of competing classes of explanation for the origin of biological in- formation (that is, the competing explanans). ..." >> [xviii]

A quick look at the truth of Creation

<< We have laws that strongly contradict biological evolution, such as biogenesis, which states that no living thing can come from dead matter. However, evolutionists insist that somehow, at some point, bodies formed from nonliving substances. This is impossible by powerful reasons:

1. The early atmosphere had prevented the emergence of life, with or without oxygen.

2. The quantities of raw materials required to synthesize life in aqueous media are so large that they had ever brought together.

3. The proteins and deoxyribonucleic acid are too complicated to occur randomly.

4. Biological systems have to work together, and do something that is of benefit to the organism. Orchestration is too specialized.

5. The living cell requires that all work at the same time, or nothing works and there is no life. Life cannot produce . Only transmitted from one living thing to another.

6. Ultimately, life can only be created by an external agent to direct and control intelligent processes.

If this does not bring God to your mind, as the only means capable of creating life, you may not be very active, or have an ulterior motive for not seeing it.

All Nature expresses a design. Never will we find a design without a designer. There are from constantly moving atoms to large star systems that were set up by someone who knew exactly what he was doing. It is so accurate that is statistically impossible to produce randomly.

Inheritance laws, restrictions to biological variations, the inability to reproduce the true hybrid, the lack of consistency in natural selection, the fate of the mutations, the absence of gradual changes, altruism, genetic codes and human language ... this gives colossal evidence against a blind and undirected process like evolution.

Why the evolutionary process did stop? has no response from evolutionists. We do know: did not stop anything because there has never been evolution in process at any time in the history of the universe.

The arguments that evolution has tried to support their case are mostly misapplied concepts, misconceptions or distorted information. Each of their proposals collapses with a thorough inspection.

The complexity of life baffles evolutionists, and they encounter situations that cannot be explained. Their attempts to unravel these mysteries cause more questions than answers, and finally babble: "Somehow it happened, didn't?" or their favorite "But life exists ..."

If evolution falls short in explaining life on this planet, it cannot nor try to propose the origins of the universe. Not even offers a plausible explanation for the formation of the moon, much less of the planets in our solar system.

Now, the natural laws that govern the universe, such as thermodynamics, placed an obstacle insurmountable in the way of evolution. The Big Bang is scientifically absurd, but they hold it because they have no other choice but God, and that is the one they unreasonably refuse.

But when we look around us with eyes wide open, we realize that there are many clues that point to a recent and wise creation, that is, with an age that is no more than 10,000 years.

The amount of salts dissolved in the oceans, the Earth's magnetic field, the rate of erosion, and many other features of our planet and even space, betray the youth age of the universe.

Evolutionists have tenfold multiplied the age of the earth every twenty years since the beginning of the century, so it now becomes 100,000 times older than it was in 1900.
Of course, before the cell was thought as complex as a ping-pong, but discovering its specialized parts ... the only thing that occurs to them is to give more time to the process.

But not even a ping-pong ball can be created by itself; and still if it did, will not come to life, no matter with eternity available. The time factor is not the magic ingredient that evolution needs to be feasible.

Fossils refuse to cooperate with the theory of evolution. They carry severe problems for scientists who believe in it. Discontinuities are impossible to fill, the mass graves speak of rapid and violent burial, and not a deposit of millimeters of ground per decade over corpses exposed to the weather.

The geologic column used as reference, is imaginary.

Fossils have been inopportunely found to shatter the evolutionary order so carefully made.

It comes out the process of setting dates, evolutionary style: **circular reasoning**, in which a fossil date rock, and the rock date the fossil. How old, really? Unknown. How old, guessed? Millions of years, of course.

Dating methods are frauds; none is absolute. All are based on unverifiable assumptions, and when they score a date that is at odds with the evolutionary idea, it arbitrarily is given   another. The radiometric methods (potassium-argon and rubidium-strontium) give dates excessively   old, with margins of error of up to 2 or 3 million years .

The carbon-14 method does not work either. First, it can only be used in organic materials, and has a maximum range of 30,000 years. However, carbon-14 has not yet reached its equilibrium; so actually all the dates that have been obtained with do not exceed 5000 years. Interesting, or not?

This method has set dates to 3000 years old in living animal species. Imagine what it will do with older remains.

Therefore, no date provided by the common methods is reliable, and they will do well in doubting all things dated back hundreds of thousands of years, and even worse, millions of years… >>[xix]

Blaise Pascal (1623-1662), Mathematician, Physicist and French Philosopher, said:

"We are generally the better persuaded by the reasons we discover ourselves than by those given to us by others."

"Man is equally incapable of seeing the nothingness from which he emerges and the infinity in which he is engulfed."

Arno Penzias (1933-present), eminent American physicist, Nobel prize in Physics in 1978, said:

"Astronomy leads us to a unique event, a universe which was created out of nothing, one with the very delicate balance needed to provide exactly the conditions required to permit life, and one which has an underlying (one might say 'supernatural') plan."

After having many years inquiring on the subject of evolution and the origin of life, I can clearly declare:

" THERE DOES NOT EXIST a person who has studied SCIENCES and is in his/her healthy judgment, that can sincerely say, that he/she believes everything that exists in the universe arose from nothing, by pure chance " (JR)

Many atheistic scientists, willing to deny the creation because they consider it to be a fable, prefer believing their own invented FABLES, putting aside their capacity to reason.

Now, I am going to put on display some fables created by those with the ability to see, but they do not want to accept what they observe ...

FABLES THEY TEACH IN THE SCHOOLS.

The Fable Without Comparison: HISTORY OF THE BEGINNING OF THE EARTH, According to The Atheistic Scientists.

(REMEMBER: THEY "DO NOT BELIEVE IN MIRACLES")

(Try to be aware of the great amount of "miracles" that should have happened)

The Origin and Evolution of Life

<< *"Think about how you rewind a videotape on a VCR. Then imagine "rewinding" the universe. As you do this, the galaxies star moving back together. After 12 to 15 billion years of rewinding, all galaxies, all matter, and all of space are compressed into a hot, dense volume about the size of the sun. You have arrived at time zero".*

"That incredibly hot, dense state lasted only for an instant. What happened next is known as the big bang; a stupendous, nearly instantaneous distribution of matter and energy throughout the universe. About a minute later, temperatures dropped a billion degrees".

"Fusion reactions created most of the light elements, including helium, which are still the most abundant elements in the universe. Radio telescopes can detect cooled, diluted background radiation – a relic of the big bang – left over from the beginning of time".

"Over the next billion years, uncountable numbers of gaseous particles collided and condensed under gravity's force to become the first stars. When the stars were massive enough, nuclear reactions ignited inside them and gave off tremendous light and heat. Massive stars

continued to contract, and many became dense enough to promote the formation of heavier elements".

"All stars have a life history, from birth to an often spectacularly explosive death. In what might be called the original stardust memories, the heavier elements released during the explosions became swept up in the gravitational contraction If new stars. They became the raw materials for the formation of even heavier elements…"

"… Now imagine a time long ago, when explosions of dying stars ripped through our galaxy and left behind a dense cloud of dust and gas that extended trillions of kilometers in space. As the cloud cooled, countless bits of matter gravitated toward one another. By 4.6 billion years ago, the cloud had flattened into a slowly rotating disk. At the dense, hot center of that disk, the shining star of our solar system –the sun- was born…"

Origin of the Earth

"…clouds are mostly hydrogen gas, along with water, iron, silicates, hydrogen cyanide, ammonia, methane, formaldehyde, and other small inorganic and organic substances."

"The contracting cloud that became our solar system probably was similar in composition.
The cloud's edges cooled between 4.6 and 4.5 billion years ago. Electrostatic attractions and gravity's pull caused mineral grains and ice orbiting the new sun to start clumping together. In time, larger, faster clumps collided and shattered. Some grew more massive by sweeping up asteroids, meteorites, and other rocky remnants of collisions. They evolved into planets."

"While the Earth formed, heat generated by asteroid impacts, internal compression, and radioactive decay of minerals melted much of its rocky interior."

"Molten nickel, iron, and other heavy materials moved into the interior; lighter ones floated to the surface. The process resulted in a crust, mantle, and core. The crust became an outer zone of basalt, granite, and other low-density rocks. It rested on a zone of intermediate-density rocks, the mantle. In turn, the mantle enveloped an immense core of high-density, partially molten nickel and iron".

"Four billion years ago, the Earth was a thin-crusted inferno (Figure 20.3a). Less than 200 million years later, life appeared on its surface! We have no record of its origin, probably because movements in the mantle and crust, volcanic activity, and erosion obliterated all traces of it…"

"Hot gases blanketed the Earth when the first patches crust formed. We suspect that this first atmosphere was a mix of gaseous hydrogen (H_2), nitrogen (N_2), carbon monoxide (CO), and carbon dioxide (CO_2). Did it hold gaseous oxygen (O_2)? Probably not.

Rocks subjected to intense heat, as happens during volcanic eruptions do release oxygen, but not much. Also, free oxygen would have reacted at once with other elements. Remember, oxygen has an electron vacancy in its outermost shell and it tends to bond with other atoms."

"If the early atmosphere had not been relatively free of oxygen, organic compounds necessary to assemble, cells would not have been able to form on their own spontaneously. Any oxygen would have attacked their structure and disrupted their functioning".

"What about water? Although dense clouds cloaked the early Earth, any water falling on the molten surface must have evaporated at once. In time, the crust cooled and grew solid. For millions of years, rainfall and runoff eroded mineral salts from rocks. The salt-laden waters collected in crustal depressions, forming the first seas".

"If liquid water had not accumulated, membranes -which take on their bi-layer organization only in water - could not have formed. No membrane, no cell."

"Life at its most basic level is the cell, which has a capacity
survive and reproduce on its own..."

"If the Earth had condensed into a planet of smaller diameter, its gravitational mass would not have been great enough to hold on to an atmosphere. If it had settled into an orbit closer to the sun, water would have evaporated from its hot surface.

If the Earth's orbit had been more distant from the sun, its surface would have been colder, and water would have been locked up as Ice. Without liquid water, life as we know it never would have originated on Earth..." >>^{xx}

Another great Fable: The formation of Oceans

<< *"More than 4.5 billion years ago, the sun and its planets were taking shape from a rotating disk of ice, gas, and dust. This protosolar nebula was hotter and denser toward its center and cooler and less dense farther out. These gradients profoundly influenced the chemical composition of different regions of the early solar system, including the distribution of water."*

"Close to the nebula's center, high temperatures and pressures vaporized ice crystals and the light elements and compounds called volatiles. The action blew these

materials toward the outskirts of the nebula, leaving mainly grains of rock behind to form the inner planets".

"Farther out, debris coalesced in meteorites called carbonaceous chondrites, which carry up to 10 percent of their mass in ice. The giant outer planets, such as Saturn and Jupiter, which arose in this neighborhood also contain some ice. Beyond these planets, water condensed in large quantities and formed comets, which are about half ice."

"Compared with these icy objects, Earth contains little water. Only about 0.02 percent of its mass is in its oceans, and somewhat more water sits beneath the surface. Nevertheless, Earth has substantially more water than scientists would expect to find at a mere 93 million miles from the sun. How did Earth come to possess its seas?"

"Over the years, planetary scientists have proposed several possible answers to that question, but until recently they've had little data for testing their hypotheses. As research in the field progresses, however, the picture is getting more complicated—not less."

"Analyses of the geochemical properties of various bodies in the solar system and computer modeling of the dynamics of ancient planetary interactions have undermined a formerly popular theory, which attributes Earth's water to a bombardment by comets late in the planet's formation."

"New hypotheses are emerging as that theory's plausibility fades, and planetary scientists are struggling to reconcile data with these alternative scenarios. There's one thing on which most geochemists and astronomers agree: The celestial pantry is now empty of a key ingredient in the recipe for Earth."

"Because comets contain a greater proportion of water than other known celestial objects do, they make natural candidates as a source of Earth's rivers, lakes, and oceans."

"The distribution of hydrogen and water beneath Earth's surface suggests to many geochemists that water hasn't mixed deep into the planet, so they thought that the cometary bombardment applied a veneer of water to the dry planet relatively late in its formative period."

"One attraction of this late-veneer scenario has been that it fits well with the early movements of planets and the many comets in the outer solar system, says Armand H. Delsemme, an astrophysicist now retired from the University of Toledo in Ohio.

As Jupiter formed, its growing gravitational tug would have sent many icy comets hurtling from the range of the giant planets to all reaches of the solar system."

"Over a billion years, at least hundreds of millions of comets collided with Earth, Delsemme says. The bombardment would have been especially heavy just after Earth formed".

"Attributing water on Earth to these latecomer comets neatly explains a couple of things: first, how water that originated at the outer edges of the solar system got to at least one of its inner planets, and second, how water arrived late enough in Earth's formation for the planet to have sufficient gravity to retain it…" >>[xxi]

Some FABULOUS recent news:

News: An Asteroid Killed the Dinosaurs, After All

<< Although any *T. Rex*–enthralled kid will tell you that a gigantic asteroid wiped the dinosaurs off the planet,

scientists have always regarded this impact theory as a hypothesis subject to revision based on further evidence gathered from around the globe.

Other possible causes, such as volcanism and smaller, multiple asteroid strikes, never actually went away, and over the years researchers raised important points that did not fully jibe with a history-changing celestial impact near the Yucatan peninsula one awful day some 65.5 million years ago.

A group of 41 researchers have pored over the evidence and decided that—in accordance with the original postulation put forth 30 years ago by a team led by father and son researchers Luis and Walter Alvarez—it was, indeed, a massive asteroid that slammed into Earth, creating Chicxulub Crater on Mexico's Gulf Coast that killed off many of the species on the planet, including the non-avian   dinosaurs.

The review, published online March 4 in *Science*, evaluated the whole picture, according to Kirk Johnson of the Research and Collections Division at the Denver Museum of Nature and Science and co-author of the paper. And that meant assessing the other theories that have been put forth about what spelled death for the dinosaurs.

The researchers dismiss the theory that the volcanism that produced the great Deccan Trap formation in western India at the end of the Cretaceous period might have produced enough sulfur and carbon dioxide to initiate a massive shift in climate. They note that pinpointing the times when the heavy volcanism occurred is sketchy, and it likely kicked off some 400,000 years before the extinction event.

In fact, as Johnson noted in a March 3 conference call with reporters, the emissions from these volcanoes likely warmed the planet slightly, actually making life easier for many animals and encouraging diversification and dispersion over wider geographical areas.

Some scientists have pointed to multiple layers of impact residue as evidence that there was more than one asteroid involved in generating the extinction. This theory did not seem to measure up, either.

Johnson says they see "no evidence for multiple impacts," and sites that had turned up these various layers were so close to Chicxulub itself that the chaotic event likely churned the layers into different locations in the sediment.

An assertion that the impact occurred hundreds of thousands of years before the extinctions also failed to hold water with the researchers. Evidence of the Cretaceous period shells on top of the impact crater are likely not a sign that the animals persisted after the impact, but rather that they got "washed into the hole," Johnson noted... >>^{xxii}

News: Dinosaurs "Breaking Wind" May Have Warmed Ancient Earth

<< The greenhouse gas methane produced by all sauropods across the globe would have been about 520 million tons per year. Between the heft of dinosaurs and their fiber-rich diets, it's likely they produced a lot of gas. Gassy dinos could have been a major factor in Earth's warming.

We might want to rename the *Brachiosaurus* with the moniker *Gassiosaurus*, new research indicates. The gassy emissions from these giant dinosaurs may have been enough to warm the Earth, the researchers say.

Sauropods are large plant-eating dinosaurs typified by such titans as *Apatosaurus* (once known as *Brontosaurus*) and *Brachiosaurus*. When they lived, during the Mesozoic era — from about 250 million years ago until the dinosaurs died out 65 million years ago — the climate was warm and wet. Nothing on Earth today compares with these giants.

The researchers found that the greenhouse gas methane produced by all sauropods across the globe would have been about 520 million tons per year, a number on par with the total amount of methane currently produced by both natural and man-made sources. *[Album: World's Biggest Beasts]*

The researchers, led by David Wilkinson of Liverpool John Moores University in the United Kingdom, did their best to get an accurate estimate of how much gas these big dinosaurs would have created, but their answers are still just estimates based on multiple assumptions, they warn.

The greenhouse gas methane is a natural byproduct of the digestive process of plant eaters, especially in herbivores called ruminants (like cows and camels). The researchers suspect that like ruminants, sauropods would have harbored methane-producing bacteria in their intestines to help digest these fibrous foods.

There is currently no way to tell what kind of bacteria lived in the digestive systems of dinosaurs, what gasses they produced, or what those digestive systems would have looked like, but Wilkinson thinks they would have produced methane like today's animals.

"To process that amount of vegetation they have to be relying on microbes in their digestive system," Wilkinson told LiveScience. "But without a time machine you can't be sure..." >>[xxiii]

News: Meteorites pummeled our planet with gold

Your earrings, your wedding ring, the platinum in your computer or phone – all of them might be alien artifacts... According to a team led by Matthias Willbold of the University of Bristol, U.K., gold, platinum, and other precious metals were brought to Earth by a massive meteor shower.

What happens is that certain metals, such as the gold, platinum, nickel, tungsten and iridium are attracted to iron; iron is the main ingredient in Earth's core, so when our planet was still a molten mass, these elements should have migrated to Earth's center, thus leaving our planet stripped of almost all its precious metals. However, the crust is riddled with them – how can this be?

Geologists have long theorized that Earth was bombarded with meteorites between 3.8 and 4 billion years ago, studding the early crust with our favorite shiny metals. These metals then became incorporated into the modern mantle over time.

The idea seems to be backed up by the existence of craters on the Moon, which date back to the same time, suggesting that the satellite was also hit by this slew of meteorites. Now, even more research seems to back this theory up.

Matthias Willbold and his team sampled ancient rocks from southwest Greenland that formed some of the Earth's earliest crust, earlier than the bombardment and compared them with newer rocks from other places representing the modern mantle. They found distinct differences in concentrations of tungsten isotopes.

"This is a sort of a time capsule that gave us the possibility to calculate how much material had to be added to the Earth to satisfy the tungsten isotopic composition that we find in the Earth today," Willbold said. "It is so far the best isotopic or geochemical evidence that late bombardment ever happened," he added.

This is indeed a major leap forward, and was applauded by other scientists as well.

"Our ability to measure these (isotopes) precisely enough to see these differences is just opening a totally new window into early planet formation," said Richard Carlson of the Carnegie Institution of Washington who has also studied early Earth using isotopes.
According to their calculations, about half percent of the material in the mantle was added by those meteorites.
"That sounds like not really much, but it's about 300 billion billion tons of material," Willbold said. "All the precious metals that we find today — and probably also water — have been introduced to the accessible Earth from these late stage meteorites".

However, these events were without a doubt devastating for the planet at that time, possibly delaying life by hundreds of millions of years. However, another possibility is that it created the conditions necessary for life to appear.

Carlson explained the early bombardment of meteorites would have been "a terrible event for life. It probably would have melted the planet, blown off any existing atmosphere.", but he also adds that:

"It's possible that this late veneer brought in the goodies afterward and it brought them in gently enough that they stuck around. I don't know that it brought in life, but it brought in maybe the constituents of it: the water, the right kind of surface temperatures, and the atmosphere that is conducive to life…" >>[xxiv]

News: Our universe could have formed earlier speculation of another Universe

<< A mathematical model provides a new theory on the formation of galaxies, stars and planets

Our Universe did not originated in a big bang, but formed from another previous twin universe to ours, as a mathematical model that provides a new theory on the formation of galaxies, stars and planets. That other twin universe would be like a mirror image of the present, as the two follow the same dynamic equations, contain the same amount of matter and follow the same evolution.

But the twin, unlike ours, is contracting, so it would be like seeing our own universe walk back in time and not all would be equal in both (e.g. people and their stories). The model suggests that our universe in turn generate other similar universes that expand while ours shrinks.

Our Universe may be the result of a Big Bounce occurred in a previous universe much like our own, instead of being led by a Big Bang, says a team of physicists from Mexico and Canada.

Up recently, scientists were not raised what could have existed before the Big Bang(literally "big bang"), which describes the development theory of the early universe and its form. According to this theory, the universe began expanding from a material point of infinite energy density and, at one point, exploded in all directions leading to the Universe in which we exist today.

Nevertheless, in recent years, an alternative hypothesis on the origin of the universe even more striking and interesting is emerging, at least from the point of view of its novelty, which proposes that our universe emerged from the collapse of another previous universe very much like ours, which would mean that our Universe is the son of another Universe.

Twin Universe

This hypothesis is included within the theory LQG (Loop Quantum Gravidity or loop quantum gravity), and suggests the possibility that before the Big Bang there was a Big-Bounce (literally, a great rebound) in a universe before ours, and that "big bounce" would have resulted in the emergence of our universe.

As magazine PhysOrg explains, physicists Alejandro Corichi , of the National Autonomous University of Mexico, and Parampreet Singh , from Perimeter Institute for Theoretical Physics of Ontario (Canada), have found their appearance through the development of a simplified LQG model.

According to Singh's statements to PhysOrg, "the importance of this concept is that it gives us an answer to what happened to the universe before the Big Bang". Singh added that the study also shows that the other universe was very similar to ours.

Cosmic Amnesia

This finding rests on previous research. Last year, a professor of physics at Penn State University in the United States called Martin Bojowald published an article in the journal Nature Physics, explaining the development of a simple mathematical model (a mathematical time machine, as University of Pennsylvania reported) allowing integration of the General Theory of Relativity of Einstein and some equations of quantum physics, and composing the first mathematical description of the existence of the Big Bounce.

This description revealed a universe before ours, which contracted before the Big Bounce and finally gave birth to our expanding universe. Bojowald also came to a further conclusion: those successive universes would not be perfect replicas of each other.

Through the creation of the mathematical model of Bojowald, no observation of our universe had been able to so far understand the state of that other pre-bounce universe, since apparently nothing was left of it after the phenomenon that produced our universe. Bojowald described it as "cosmic amnesia."

Twins in Time and Laws

Corichi and Singh seem to have overcome that amnesia. Modifying LQG theory with the inclusion of a key equation called the quantum constraint (generating sLQG version of this theory), they have been able to show that the relative fluctuations of volume and momentum belonging to the pre-bounce universe (pre-bounce universe) were kept at either side of the rebound.

The conclusion the physicists arrive at is that the other twin universe would have the same physical laws and the same temporary notion as ours. In fact, "seen from afar, both universes could not be distinguished from each other," said Singh in PhysOrg.

Our current universe, about 13,700 million years after the Big Bounce, would share many features with the previous Universe when he was the age of 13,700 million years before the bounce. In a sense, our universe and its twin would be mirror images of one another, with the time frame as the Big Bounce symmetry line.

Both universes would appear, for example, that the two would follow the same dynamic equations, and contain the same amount of matter, and follow the same evolutionary cycle. But the twin, unlike our universe, is contracting; so it would be like seeing our own universe walk back in time

Universal Reproduction

But, as the physicists' state, not everything is equal in each universe with respect to the other. For example, the existence of the other twin universe in relation to ours would not mean there are exact replicas of each person, or people who have lived our lives exist in the alternate reality.
According to Singh, it would be something like what happens in human twins: when studied by scale, you can see even small differences between them, as fingerprints or DNA.

"In addition, other factors of the twin universe must still be clarified." said the scientist. Most important: did similar properties survive in case that, instead of applying a simplified model, more complex variables are introduced as the possibility of traces of previous galaxies before the most recent ones. Is there a possibility of similar structures in these galaxies of an expanding universe?

Finally, Corichi and Singh's model could be used to explore the future of our own universe. It is indeed possible that a generalization of the model established by physicists predict a Big Bounce of our own universe. Thus, it is possible that our universe may in turn generate other universes, and that these resemble each other.

The scientists will soon make public their research in the journal Physical Review Letters, but have anticipated the text in Arxiv. >>[xxv]

A commentary about Antimatter, Dark matter, Black holes and parallel universes

<< Antimatter

According to observations made in particle accelerators, for every particle of matter there is a contrary one called antiparticle. Then the Big Bang should have created at least as much matter as antimatter, which should exist as anti planets, anti stars and anti galaxies.

Antiparticles have the same mass than particles with charges corresponding but inversely, i.e. if matter is positively charged, the antimatter is negatively charged and vice versa. If matter and antimatter collide, there would be an explosion of unimaginable magnitude.

Dark matter

It is estimated that the visible matter accounts for 10 percent of all matter in the universe. Although, a priori, this statement seems to lack logic, and is based on years of mathematical calculations that determine the "necessity" of their presence.

There is presently no visible evidence of the existence of dark matter, but this is necessary to explain many events that occur in the universe. This explains the adequacy of gravitational forces, for example, galaxies rotate faster than they should based on their visible mass, and galaxy clusters would need 90 percent more mass to stay together. In recent years, there have been numerous scientific studies looking for antimatter in space, discovering some revealing data.

According to this theory, it is speculated the existence of one, several or infinite parallel universes of dark matter, twin universes where the space-time concept is not necessarily as we know it in our world.

Black holes and parallel universes.

The stars with a mass higher than 8 times that of the Sun, end their lives in a huge explosion (supernova). The result of this explosion is a very unstable mass. Most of the matter in the star is thrown into space in the blast but the center shrinks because of the immense gravitational force, until it contracts to a size less than a small asteroid, but with a mass such that a spoonful would weigh thousands of millions of tons. The more mass there is, the stronger the gravity, and in turn, a greater amount of gravity pulls more amount of mass.

This happens in geometrical progression up to the paradox of a dimensionless point with a gravitational force which even light cannot escape. We could then say that "the vacuum cleaner has swallowed itself" and has created a black hole. Black holes emit X-rays and, therefore, can be detected, i.e., meaning X-ray source which has no light is detected.

These holes cause a distortion in the universe, as if we placed a powerful vacuum ten inches deep and with the mouth up, in the middle of the sea. The water would be transported from the surface to the bottom creating a whirlpool.

But ...Where does the matter and light swallowed by the black hole go? To date, no one has been able to answer this question...

...For now, it is not possible to apply the theory of relativity to anything "small" or quantum theory to anything "big", so huge gaps appear when trying to make predictions with respect to dark matter or black holes.

Until the "unification theory" arrives, a final theory that unifies general relativity with quantum physics, we cannot move forward. For now we can only imagine ... It appears that, for now, the Hermetic Principle of Correspondence is not provable: "As above, so below ..." >>xxvi

These are opinions of some scientists:

C. J. Isham (1944 - present), astrophysicist, said:

"Probably the best argument ... that the Big Bang supports theism is received with obvious dissent by some atheistic physicists. To the present this has lead to scientific ideas ... advancing with a tenacity that much exceed its intrinsic value, so one can suspect the operation of psychological forces that lie much deeper than the usual academic desires of a theoretician to support its theory."

Paul Davies (1946 – present), physicist, said:

"It is rather hard to resist the impression that the present structure of the universe, apparently so sensitive to minor alterations in numbers, has been rather carefully thought out... The seemingly miraculous concurrence of these numerical values must remain the most compelling evidence for cosmic design."

George Greenstein (1940-present), astronomer, said:

"As we survey all the evidence, the thought insistently arises that some supernatural agency—or, rather, Agency—must be involved. Is it possible that suddenly, without intending to, we have stumbled upon scientific proof of the existence of a Supreme Being? Was it God who stepped in and so providentially crafted the cosmos for our benefit?"

What is YOUR opinion?

What an ANT teaches...

Did you know...

... AN **ANT**

REASONS

MORE THAN

AN **ATHEIST**?

No wonder, GOD
Describes as:

"FOOLISH"

To The Atheists
(Psalm 14:1)

APENDIX 1

Concepts of Science, Scientific Knowledge and Logic

Definition of Science

<< From the origins of mankind our species has eagerly pursued knowledge, trying to categorize and define clear and well differentiated concepts from each other. In ancient Greece, scholars decided to establish a concept that would encompass all knowledge, science.

It is first necessary to define knowledge as a body of information acquired through experience or introspection and can also be organized in a structure of objective facts accessible to different observers. That set of techniques and methods used to achieve such knowledge is considered "science". The word comes from the Latin *scientia* and precisely means knowledge.

The systematic application of these methods creates new objective knowledge (scientific), and it acquires a specific shape.. First, a hypothesis is given, and it is tested through scientific methodology and subjected to quantification.

Moreover, these predictions of science can be located within a structure by the detection of universal rules that describe how a system works. These are the same universal laws that allow us to know in advance how the system in question will act under certain circumstances. (Not-rules that can describe)

Science can be divided into basic science and applied science (when applying scientific knowledge to human needs). >> [xxvii]

Value of Science

<< The views concerning the value of science are varied and even opposing.

For some, the role of science is to give a *possible explanation* of the facts. If science explains satisfactorily in accord with reason, then the theories presented and postulated are valid.

For others, science offers a unique system to decipher our reality, which is also unique. There cannot exist two realities, so two valid explanations of reality cannot co-exist. Science and reality are one and the same. For these people the role of science is *cognitive*, seeking to know reality.

Others claim that science is a *creation* of man. The main value of science lies in discovering the harmonies in thought, which may or may not coincide with the harmony of reality. Many mathematicians saw their science as a game of chess, where thinking dictates the laws which "thought" then subjects itself to. The role of science, thus understood, is primarily *aesthetic*.

Still others argue that the role of science is *practical*: science is a tool to master reality.

Objectivity of Science

Individualistic thought, preferences, tendencies, or aspirations should not intervene, nor add to, the explanation of facts. Though the man of science may be driven by passion, and be satisfied with the results, knowledge itself should not be affected by these elements. One can say that the pursuit of knowledge is an act of courage because any interest other than the quest for truth must be sacrificed.

The scientist works with his intelligence and self interest and sentiment are subservient to this search. He does not see it necessary to use intelligence to manipulate facts in order to subject them for purposes other than the pursuit of truth.

Characteristics of Scientific Knowledge

Scientific knowledge is a critical knowledge (fundamental), methodical, verifiable, consistent, unified, organized, universal, objective, communicable (through scientific language), rational, provisional and explains and predicts facts by using laws.

Scientific knowledge is *critical* because it tries to distinguish the true from the false. It achieves this by justifying its body of knowledge and continually proving its truths. In demonstrating the correctness of its foundation we can accept its conclusions as being right.

Its foundation is established *through the methods* of research and testing, as the researcher follows procedures and develops the task based on a previous plan. Scientific research is not erratic but is planned.

Verification is possible by passing the test of prior results. Verification techniques evolve over time.

It is *systematic* because it is an organized entity, as new knowledge is integrated into the system, interacting with the existing ones. It is *ordered* because it is not an aggregate of isolated information, but a system of ideas connected.

It is *unified* knowledge because it doesn't seek an understanding of the singular and concrete, but the knowledge of the general and abstract, that is, those things identical and permanent.

It is *universal* because it applies to all persons without recognized boundaries or determinations of any kind, and does not vary with different cultures.

It is *objective* because it is valid for all individuals and not just for one particular person. Overall value is not unique or individual value. It strives to know reality as it is and guarantee the integrity of this objectivity with its techniques and methods of research and testing.

It is *communicated* through the language of science, which is precise and unambiguous, understandable to any qualified individual with the ability to access the information needed to check the validity of the theories in logical and verifiable aspects.

It is *rational* because science knows things by using intelligence and reason.

Scientific knowledge is *provisional* because the task of science does not stop, continuing its research to better understand reality. The search for truth is an open task.

Science explains reality through *laws* that have a constant and necessary relationship to facts. There are universal propositions that establish under what conditions a certain event happens and through these particular facts unique events are understood. They also allow anticipating events, making possible the prediction of what can occur. The explanations of the facts are rational, obtained through observation and experimentation. >>[xxviii]

Definition of Logic

"The study of the logic is the study of the methods and principles used to distinguish between the correct (good) arguments and the incorrect (bad) arguments. Logic is an artificial, but formal language. It is an abstract language that wants to analyze the reasoning.

Some philosophers have defined logic as " the science that studies the formal beginning of the knowledge, that is to say, those conditions that must be fulfilled in order that a knowledge, whatever its content is, could be considered to be real and good founded, and not as a mere occurrence or as a hypothesis without any base ".

Natural Logic, and Scientific Logic

<<The capacity to think is particular and exclusive of the human being; however human thinking is not arbitrary, but subject to a string of rules or laws. In other words, his/her rational capacity could provide understanding and true knowledge; but first must be adjusted to a series of rules or laws that are, indeed, those logic care about.

The object, purpose and use of logic is to ensure the correctness of reasoning. Now, somebody could argue - and would not be erroneous- it is not necessary to have studied logic, to reason correctly. This assessment requires us to distinguish between the concepts of "natural logic" and "scientific logic":

There is a natural or spontaneous logic, prior to any culture, (we might call it common sense) enough for daily life and even for the development of different branch of learning. As a result, lawyer, journalist, employer, doctor, economist, or physicist, typically begin their studies trusting their natural "act of reason", or natural logic.

However, a complete education requires to cultivate a logic-based scientific evidence, not fulfilled with natural logic or spontaneous act of reason.

Theoretically, scientific logic remove a gap in our understanding and allows us to know the basis for the rules followed by our reason. Moreover, provides our intellectual activity with practical order and maximum resolution.

To distinguish natural and spontaneous logic, reflexive logic should be called 'artificial logic'; and, in fact, it was so called in former treaties. But this term have acquired a pejorative meaning in everyday language, and has taken the term "scientific logic". It is important to reiterate that scientific logic help grows or develops natural logic or common sense, but is not a substitute for it.

First, scientific logic allows perform fast and perfectly, long and complicated reasoning, too difficult or complex to simple common sense. On the other hand, expose and identify gaps in reasoning, fallacies and feelings that common sense may suspect or consider, but is unable to reject or correct.

Scientific logic, as the study of rules and laws of thinking, is extracted from natural logic; to be precise, it enlighten what the natural logic naturally receive, and systematizes rules or draws a number of conclusions. Thus the idea of a scientific logic that initiate as a technique, and systematically develops the use of reason>> [xxix]

APENDIX 2.

Questions and Answers for Atheists

<< Dealing with atheists is actually easy to do. They don't have any evidence for their atheism, and they can't logically prove there is no God. They can only attack the Bible and attack Christians' ideas of God. But, if you listen to them, you will soon find their logic has many holes in it.
It takes practice, but you can do it.

The following statements are for copying and pasting into chat rooms. Use them to see how atheists react. Use them to learn how to respond better to atheists. Please understand that these are not "stoppers."

But, they can be challenging to atheists. Also, see how long it takes before they become condescending. Do not return their condescension. Instead, ask them to give rational reasons for their positions. In the process of interacting with them, learn how to debate with them better.

Ways to Attack Atheism

By asking questions

i.   Atheism is an intellectual position. What reasons do you have for holding that position? Your reasons are based upon logic and/or evidence or lack of it. So, is there any reason/evidence for you holding your position that you defend?

ii.   If you say that atheism needs no evidence or reason, then you are holding a position that has no evidence or rational basis? If so, then isn't that simply faith?

iii. If you say that atheism is supported by the lack of evidence for God, then it is only your opinion that there is no evidence. You cannot know all evidence for or against God; therefore you cannot say there is no evidence for God.

iv. If you say that atheism needs no evidence to support it because it is positions about the lack of something, then do you have other positions you hold based upon lack of evidence...like say, screaming blue ants? Do you hold the position that they do not exist or that you lack belief in them, too?

By using logic

v. How do you account for the laws of logic in a universe without God? The laws of logic are conceptual by nature and absolute. Being absolute, they transcend space and time. They are not the properties of the physical universe (since they are conceptual) or of people (since people contradict each other, which would mean they weren't absolute). So, how do you account for them?

a. This approach is a bit more complicated. If you use this one, first be familiar with The Christian Worldview, the Atheist Worldview, and Logic.

b. First of all, when using logic, you should be familiar with basic laws of logic and logical fallacies. It is very useful to point out the various logical fallacies to atheists as they commit them. Therefore, please be familiar with Logical Fallacies or Fallacies in Argumentation.

c. The laws of logic are conceptual by nature and are always true, all the time, everywhere. They are not physical properties. How do atheists account for them from an atheist perspective?

vi. Everything that was brought into existence was caused to exist. Can you have an infinite regression of causes? No, since to get to "now" you'd have to traverse an infinite past. It seems that there must be a single uncaused cause. Why can't that be God?

vii. Examples of logical absolutes:

a. Examples of logical absolutes are: something cannot be itself and not itself at the same time (Law of non-contradiction). A thing is what it is (Law of identity). A statement is either true or false (Law of excluded middle). These are simple, absolute logical absolutes.

viii. If atheism is true: The universe has laws. These laws cannot be violated. Life is a product of these laws and can only exists in harmony with those laws and is governed by them. Therefore, human thought, feelings, etc., are programmed responses to stimuli and the atheist cannot legitimately claim to have meaning in life.

ix. Human constructs?

a. If the laws of logic are human constructs then how can they be absolute since humans think differently and often contradictorily? If they are produced from human minds, and human minds are mutually contradictory, then how can the constructs be absolute? Therefore, the laws of logic are not human constructs.

B.   The Universe exists

i.   The universe exists. Is it eternal or did it have a beginning? It could not be eternal since that would mean that an infinite amount of time had to be crossed to get to the present. But, you cannot cross an infinite amount of time (otherwise it wouldn't be infinite). Therefore, the universe had a beginning.

ii.   Something cannot bring itself into existence. Therefore, something brought it into existence.

iii.   What brought the universe into existence? It would have to be greater than the universe and be a sufficient cause to it. The Bible promotes this sufficient cause as God. What does atheism offer instead of God? If nothing, then atheism is not able to account for our own existence.

iv.   The universe cannot be infinitely old or all useable energy would have been lost already (entropy). This has not occurred. Therefore, the universe is not infinitely old.

v.   Uncaused Cause

    a. Objection: If something cannot bring itself into existence, then God cannot exist since something had to bring God into existence. Answer: Not so. You cannot have an infinite regression of causes lest an infinity be crossed (which cannot happen). Therefore, there must be a single uncaused, cause.

    b. All things that came into existence were caused to exist. You cannot have an infinite regression of causes (otherwise an infinity of time has been crossed which is impossible because an infinity cannot be crossed).

Therefore, logically, there must be a single uncaused cause that did not come into existence.

Responding to Atheist Statements about God

C. "I lack belief in a God."

i. If you say that atheism is simply lack of belief in a god, then my cat is an atheist the same as the tree outside and the sidewalk out front, since they also lack faith. Therefore, your definition is insufficient.

ii. Lacking belief is a non-statement because you have been exposed to the concept of God and have made a decision to accept or reject. Therefore, you either believe there is a God, or you do not, or you are agnostic. You cannot remain in a state of "lack of belief."

iii. If you lack belief in God, then why do you go around attacking the idea of God? If you also lack belief in invisible pink unicorns, why don't you go around attacking that idea?

D. "I believe there is no God."
i. On what basis do you believe there is no God?

E. "I don't believe there is a God."
i. Why don't you believe there is no God?

F. "There is no God."

i.   You cannot logically state that there is no God because you cannot know all things so as to determine that there is no God.

G.   "There is no proof that God exists"
i.   To say "there is no proof for God's existence" is illogical because an atheist cannot know all things by which he could state that there is no proof.
He can only say he has not yet seen a convincing proof; after all, there may be one he hasn't yet seen.

H.   "All of Science has never found any evidence for God."
i.   That is a subjective statement. There are many scientists who affirm evidence for God's existence through science.

ii.   Your presupposition is that science has no evidence for God, but that is only an opinion.

iii.   Science looks at natural phenomena through measuring, weighing, seeing, etc. God, by definition, is not limited to the universe. Therefore, it would not be expected that physical detection of God would be found.

I.   "What is God?" or "Define God."
i.   God is the only Supreme Being who is unchanging, eternal, holy, and Trinitarian in nature. He alone possesses the attributes of omniscience, omnipresence, and omnipotence. He alone brought the universe into existence by the exertion of His will.

J.   "Prove your God is real."
i.   I can no more prove to you that God is real than I can prove to you that I love my family. If you are

convinced I don't love my family, no matter what I say or do will be dismissed by you as invalid. It is your presuppositions that are the problem, not whether or not God exists.

ii.   I can no more prove to you that God is real than you can prove that the universe is all that exists. Your demand of proof precludes acknowledgement of many types of evidence because your presuppositions don't allow it.

iii.   The universe exists. It is not infinitely old. If it were it would have run out of energy long ago. Therefore, it had a beginning. The universe did not bring itself into existence. Since it was brought into existence by something else, I assert that God is the one who created the universe.

   a.   When the atheist complains, ask him to logically explain the existence of the universe. Point out that opinions and guesses don't count.

Responding to Atheist Statements about the Bible

K.   "The Bible is full of contradictions."

i.   Saying the Bible is full of contradictions does not mean it is so. Can you provide a contradiction that we can examine in context? There are many websites that address alleged contradictions. Here is one: www.carm.org.

Responding to Atheist Statements about Evolution and Naturalism

L.   "Evolution is a fact."

i.   That depends on if it is micro or macro. Micro variations occur, but macro variations (speciation) have not been observed. The best we have are fossils and they have to be interpreted. Besides, there are plenty of gaps in the fossil record.

ii.   Have you read any books that discuss the contrary evidence to evolution? If not, then how can you say you are educated enough to say it is a fact?

M.   "Naturalism is true; therefore, there is no need for God."

i.   Naturalism is the belief that all phenomena can be explained in terms of natural causes and laws. If all things were explainable through natural laws, it does not mean God does not exist since God is, by definition, outside of natural laws since He is the creator of them.

Responding to Atheist Statements about Truth

N.   "There are no absolute truths."

i.   To say there are no absolute truths is an attempt to state an absolute truth. If your statement is true, then it is self-contradictory and not true, and you are wrong. >>^{xxx}

APENDIX 3

RECENT NEWS OF THE PROPAGANDA BLITZ FROM MILITANT ATHEISTS

Although atheism is devoid of any foundation worthy of anyone believing its anti-philosophical and anti-scientific pretensions, atheists continue to exert pressure on society with their many snares and presumptions. Their goal is to target and make an impression on those most susceptible and ingenuous.

The following pages contain news items published on this matter.

National Atheist Ad Campaign Targets Black Community

By Jeff Schapiro, Christian Post Reporter. *February 1, 2012 | 4:50 pm*

For complete reading of this article, please go to:
http://www.christianpost.com/news/national-atheist-ad-campaign-targets-black-community-68442/

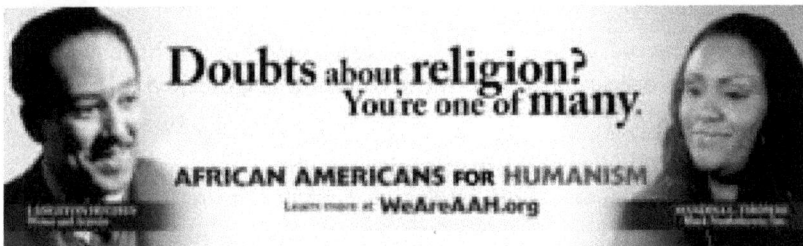

Atheist organizations from around the country have taken to billboard advertising to promote their views and their organizations over the last few months, but a new campaign by one atheist organization is focusing on reaching one group of people in particular: African-Americans. "A lot of people think there's one black experience. A lot of people think that if someone's black it means that they're religious. So we want to be able to show people that that's not true, that there are non-religious people out there," Debbie Goddard, director of African Americans for Humanism (AAH), told The Christian Post on Wednesday.

The AAH launched an advertising campaign in late January in six major U.S. cities – New York City, Los Angeles, Chicago, Atlanta, Washington, D.C. and Durham, North Carolina – with a seventh city, Dallas, being added on Feb. 6. The campaign was designed to coincide with February's Black History Month.

Each billboard, poster or banner that goes up says "Doubts about religion? You're one of many" and has AAH's website printed on it. Each sign will also feature the image of a famous historic black freethinker – like poet Langston Hughes, social reformer Frederick Douglass or writer Zora Neale Hurston – across from the photo of a contemporary black atheist leader.

Atheist Ads on Buses Rattle Fort Worth

Joyce Marshall/Fort Worth Star-Telegram

Why the bus ads now? "It can be pretty lonely for a nonbeliever at Christmastime around here," the head of an atheist group says.
By JAMES C. McKINLEY Jr.
Published: December 13, 2010

For complete reading of this article, please go to:
http://www.nytimes.com/2010/12/14/us/14atheist.html?_r=2ref=religion_and_beli
ef&pagewanted=all

FORT WORTH — Stand on a corner in this city and you might get a case of theological whiplash.

A public bus rolls by with an <u>atheist</u> message on its side: "Millions of people are good without God." Seconds later, a van follows bearing a riposte: "I still love you. — God," with another line that says, "2.1 billion Christians are good with God."

A clash of beliefs has rattled this city ever since atheists bought ad space on four city buses to reach out to nonbelievers who might feel isolated during the Christmas season. After all, Fort Worth is a place where residents commonly ask people they have just met where they worship and many encounters end with, "Have a blessed day."

"We want to tell people they are not alone," said Terry McDonald, the chairman of Metroplex Atheists, part of the Dallas-Fort Worth Coalition of Reason, which paid for the atheist ads. "People don't realize there are other atheists. All you hear around here is, 'Where do you go to church?' " But the reaction from believers has been harsher than anyone in the nonbeliever's club expected. Some ministers organized a boycott of the buses, with limited success. Other clergy members are pressing the Fort Worth Transportation Authority to ban all religious advertising on public buses. And a group of local businessmen paid for the van with the Christian message to follow the atheist-messaged buses around town.

Atheists want sign honoring 9-11 firefighters removed

For complete reading of this article, please go to:
http://radio.foxnews.com/2011/06/21/atheists-want-sign-honoring-9-11-firefighters-removed/

A group of New York City atheists is demanding that the city remove a street sign honoring seven firefighters killed in the Sept. 11, 2001 terrorist attacks because they said the sign violates the separation of church and state.

The street, "Seven in Heaven Way," was officially dedicated last weekend in Brooklyn outside the firehouse where the firefighters once served. The ceremony was attended by dozens of firefighters, city leaders and widows of the fallen men.

"There should be no signage or displays of religious nature in the public domain," said Ken Bronstein, president of New York City Atheists. "It's really insulting to us."

Bronstein told Fox News Radio that his organization was especially concerned with the use of the word "heaven."

"We've concluded as atheists there is no heaven and there's no hell," he said. "And it's a totally religious statement. It's a question of separation of church and state." He was nonplussed over how his opposition to the street sign might be perceived – especially since the sign is honoring fallen heroes. "It's irrelevant who it's for," Bronstein said. "We think this is a very bad thing."

David Silverman, president of American Atheists, agreed and called on the city to remove the sign.

"It implies that heaven actually exists," Silverman told Fox News Radio. "People died in 9-11 but they were all people who died, not just Christians. Heaven is a specifically Christian place. For the city to come up and say all those heroes are in heaven now, it's not appropriate."

"All memorials for fallen heroes should celebrate the diversity of our country and should be secular in nature. These heroes might have been Jews, they might have been atheists, I don't know but either way it's wrong for the city to say they're in heaven. It's preachy."

Happy 4th! Atheists proclaim 'God-LESS America'. Plan to try again with banners flown over Independence Day celebrations

Published: 07/03/2012 at 8:01 PM
by Dave Tombers

For complete reading of this article, please go to:
http://www.wnd.com/2012/07/happy-4th-atheists-proclaim-god-less-america/

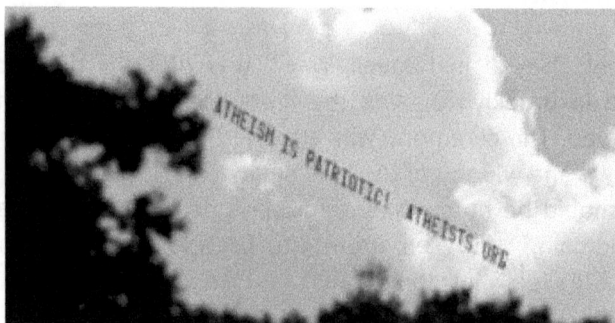

Last July 4th, WND reported on the failure of the group American Atheists to find pilots willing to fly "Godless" banners over American cities in all 50 states. Some 80 percent of pilots refused the jobs, saying things such as, "I'm not going to hell flying that sign."

According to the atheists' website, this year they aren't trying to be quite as ambitious. Instead of trying to locate pilots in all 50 states, this year the group plans to hire one pilot to fly a banner above New York City on the 4th of July. According to the group: "In celebration of the rise of atheism in America, American Atheists flew aerial banners across the country on July 4th in 2011. ... The banners proudly stated, 'God-LESS America' and 'Atheism Is Patriotic.' The banners brought the breadth and patriotism of the movement into America's conversation." However, when so many pilots the group tried to hire refused the work, the atheists cited the need to keep up their efforts. "Originally, we had planned on flying banners in all 50 states, representing the fact that atheism is the fastest growing segment in all 50 states, but we were unable to find pilots in many states willing to fly our banners, representing a clear reminder of the work we have to do," said the group. The group seems confident they will find one pilot to help declare their message.

PA. Atheists Use Race & Slave Imagery in Billboard Against the 'Barbaric' Christian Bible

Posted on March 7, 2012 at 12:28pm by Billy Hallowell
For complete reading of this article, please go to:
http://www.theblaze.com/stories/pa-atheists-use-race-slave-imagery-in-billboard-against-the-barbaric-christian-bible/

Image Credit: American Atheists

Last month, The Blaze told you about a battle that's been brewing between atheists and Pennsylvania lawmakers after the state's House of Representatives unanimously passed a resolution calling 2012 the "Year of the Bible." At the time, a staff lawyer from the Freedom From Religion Foundation called the act "shocking."

Now, two other groups, American Atheists and Pa. Nonbelievers, are being accused of invoking racial themes after posting a Bible-inspired billboard against the designation. Their sign, which tackles the issue of slavery, was erected in an area of Harrisburg, Pennsylvania, with a large African American population. The message aimed at railing against politicians who supported the resolution, was posted just blocks away from the state capitol. It comes, as many atheist-led billboard campaigns do, with a fair amount of controversy. Only this particular message was so inflammatory that it also led to a defacing. State Rep. Thaddeus Kirkland (D-Delaware), a black legislator, was one of the many voices objecting to the now-removed billboard, which featured a shackled slave. The image of the individual in shackles appeared below the words, "Slaves, obey your masters." Kirkland supported the Bible resolution and has claimed that the billboard took the Bible out of context and that it is portrayed both racism and hatred.

Atheist Billboard 'Bad Manners,' Says Christian Research Fellow

By Michael Gryboski , Christian Post Reporter
January 25, 2012 | 12:51 pm
For complete reading of this article, please go to:
http://www.sermonaudio.com/new_details.asp?ID=33283

As a Colorado atheist group purchases space for three billboards in major cities in Colorado, one Christian research fellow refers to their efforts as "bad manners."

"They say their ad is intended to spark dialogue with people of faith on the existence of God, but you don't draw people into conversation by poking fun of the beliefs," Glenn Stanton, director for Family Formation Studies at Focus on the Family in Colorado Springs, told The Christian Post. Stanton was referring to a billboard sponsored by Boulder Atheists that states, "God is an imaginary friend; Choose reality, it will be better for all of us."

"Pew reports that 92 percent of Americans believe in God or some higher being," Stanton pointed out. "And more than 70 percent say they have a firm, confident belief in God. And this atheist group equates that very widely-held belief to a small child having an imaginary friend to play with."

But the "real bummer" about the atheist billboard, according to Stanton, was that it replaced a billboard "of our area's beautiful Royal Gorge" with a "poor graphically-challenged ad."

The Boulder Atheists' ad will be posted on three billboards in Denver and Colorado Springs.

Billboard calls Jesus 'Useless Savior'. Atheists attack faith at nominating conventions

by Dave Tombers

For complete reading of this article, please go to:
http://www.wnd.com/2012/08/billboard-calls-jesus-useless-savior/

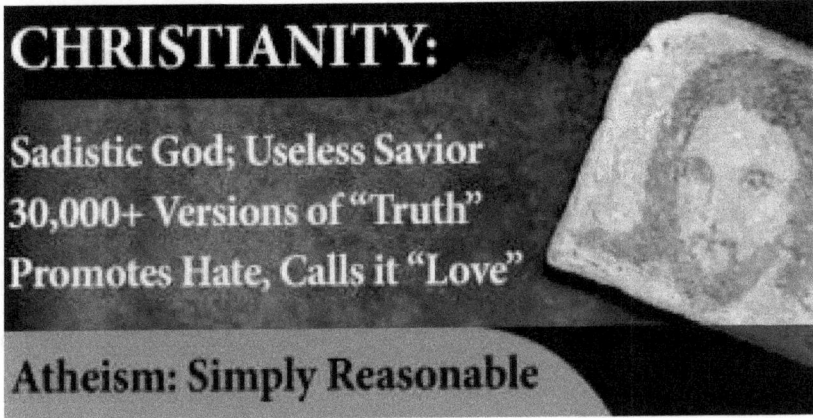

A group that pushes the philosophy of "loving yourself and your fellow man rather than a god" is attacking religious believers in a new billboard campaign in the host cities of the upcoming Republican and Democratic conventions.

American Atheists announced a billboard campaign against "the foolishness of religion in the political landscape."

According to the group, the billboards feature assertions of Christianity and Mormonism that they say have no place in politics.

One version of the campaign is scheduled to run in Charlotte, N.C., where the Democratic National Convention will be held Sept. 3-6.

The billboard, as seen on American Atheists' website, attacks Jesus as a "Useless Savior" next to an image of a piece of toast with a purported image of Jesus burned into it.

It also claims Christians have a "Sadistic God" and that Christianity has more than "30,000+ versions of truth." Christians, it says, are part of a "Hate Promoting" group that calls "Hate" "Love."

The other version of the campaign has fallen short already, as no billboard company in Tampa, Fla., the site of the Republican convention, will accept it....

Atheist display in Streator City Park stirs 1st Amendment issue: City Council will discuss public display policy

04/06/2012, 11:03 pm
Derek Barichello
For complete reading of this article, please go to:
http://mywebtimes.com/archives/ottawa/display.php?id=453539

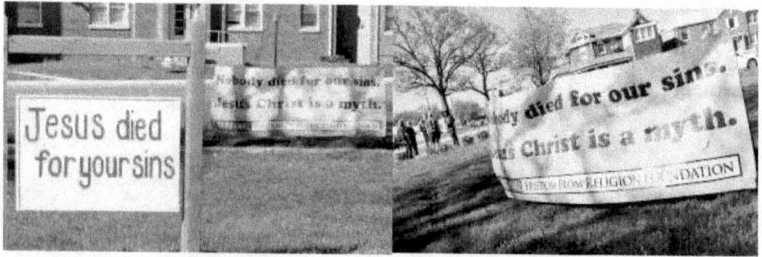

A banner placed in Streator's City Park on Thursday is now calling into question the city's policy regarding religious displays in public parks.

The Freedom From Religion Foundation out of Wisconsin placed an eight-foot by three-foot banner which reads, "Nobody died for our sins. Jesus Christ is a myth," in the northwest corner of City Park near a religious display of crosses with a sign that reads, "Jesus died for our sins."

The national church watchdog organization said it placed the banner in Streator on behalf of a local resident's request to counter the Christian display that has been on the city property since early March. City officials permitted both displays.

This is the fifth year in a row the Streator Freedom Association has displayed the Christian crosses in the park around Easter time.
"We think the city would be wise to exclude all displays from the park,"said Annie Laurie Gaylor, Freedom From Religion Foundation co-president.

Skeleton Santa Controversy at Loudoun County Courthouse

Controversial holiday display vandalized in Leesburg

Tuesday, Dec 6, 2011 | Updated 12:59 PM EST

For complete reading of this article, please go to:

http://www.washingtonpost.com/blogs/post_now/post/crucified-skeleton-santa-sparks-controversy-in-loudoun/2011/12/06/gIQAMIcxZO_blog.html

Darcy Spencer
Some residents of Loudoun County say they were outraged after someone put a display of a skeleton Santa Claus outside the Leesburg courthouse.

Christmas Controversy in Leesburg

Holiday music, holiday lights and holiday sales are unavoidable the first week of December, but tisn't really the season without a holiday display controversy in Leesburg, Va. A skeleton dressed in a Santa suit and nailed to a cross was set up on the Loudoun County courthouse lawn in Leesburg on Monday. The macabre Kris Kringle was one of the nine approved displays for this Christmas season, but it was not standing for long. Someone tore the skeleton down, sparking a debate about free speech. It's not a new argument.

In 2009 Christmas displays on the courthouse lawn were banned after the constitutionality of a Nativity scene was questioned. Last year that decision was overturned, and 10 displays were allowed on the lawn based on a first come first serve basis. Leesburg council member Ken Reid spoke out strongly against the skeletal Christmas display. "I think that it's just extremely, extremely sad," he said, "that somebody in this county who would try to basically debase Christmas like this. This really crossed the line."

117

'Good Without God': Freedom From Religion Foundation to Convene in Connecticut for Annual Atheist Convention

October 4, 2011 at 7:05pm by: Billy Hallowell

For complete reading of this article, please go to:

http://www.sequesterednews.com/news/theblaze/84882-
%E2%80%98good-without-god%E2%80%99%3A-freedom-from-religion-
foundation-to-convene-in-connecticut-for-annual-atheist-convention.html

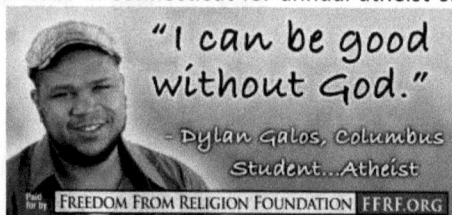

This weekend, atheists will converge in Hartford, Connecticut, to participate in the Freedom From Religion Coalition's (FFRC) annual conference. FFRC, one of the most vocal groups working to remove God from American society, wholeheartedly believes that "the most social and moral progress has been brought about by persons free from religion." The group describes its mission as follows:

The Foundation works as an umbrella for those who are free from religion and are committed to the cherished principle of separation of state and church. The 34th annual gathering of atheists will be held at the Marriott Hartford Downtown and will include a multitude of non-believing speakers. Among the individuals currently on the agenda are professors, entertainers and the like. There's Joseph Taylor, a former Christian rock band member (the band "Undercover") who is now a "nonbelieving educator." Then, there's broadway composer Charles Strouse, the lifetime atheist who wrote the musicals "Annie" and "Bye Bye Birdie." During the convention, FFRC will be honoring University of Chicago Professor Jerry Coyne with the "Emperor Has No Clothes Award." Coyne wrote a book back in 2009 called, "Why Evolution is True." Of course, these are only a few of the individuals who will be addressing conference participants (the full list can be found here). Below, watch Christopher Hitchens' acceptance speech for this same award back in 2007 (during the 30th annual event): The Guardian has more information on the event, which will likely cause some angst among the religious.

'Kids Without God': Atheist Activists Launch Shocking Web Site to Convert Kids & Teens Into Non-Believers

Posted on November 13, 2012   by Billy Hallowell**For complete reading of this article, please go to:**
http://www.theblaze.com/stories/kids-without-god-atheist-activists-launch-shocking-web-site-to-convert-kids-teens-into-non-believers/
The atheist activist community in America has taken an increasingly-active role in trying to convince citizens with doubts about their faith to fully evolve into non-believers and to "come out," publicly proclaiming their anti-theism. Think of it as a form of secular evangelism.

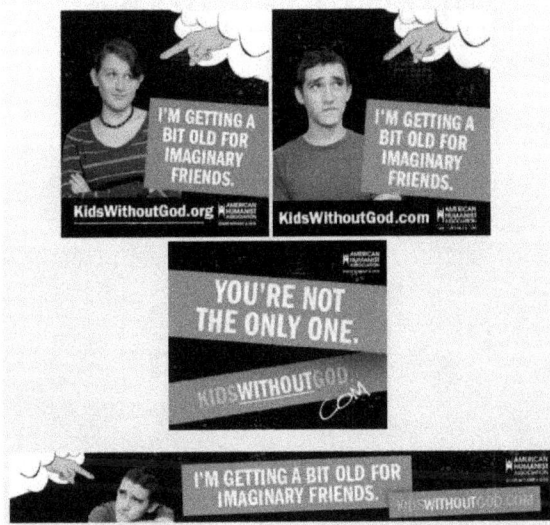

A screen shot of some of the AHA's ads for KidsWithoutGod.com

Already, non-believers have attempted to reach clergy who are in doubt through The Clergy Project. Additionally, there's a humanist church service each week in Tulsa, Oklahoma (and these are only two examples). Now, in addition to reaching adults, atheist activists have their eyes set on converting kids and teens...

119

I WOULD ASK YOU:

IF YOU KNOW THE TRUTH,

WHAT WILL YOU DO?

■ · ▬ · ▬ · ·

*"If you witness the flames of evil spreading
everywhere; extinguish them without hesitation,
with the waters of goodness"*
(JR)

BIBLIOGRAPHY

For all those that wish to abound in the subject of this book, I recommend you read:

A. Torres Queiruga. "Creo en Dios Padre". Ed. Sal Terrae, Santander, España, 1986

Albert Einstein. "Este es mi Pueblo". Editorial ElAlef. www.elaleph.com

Ángel Peña OAR. "Ateos y Judíos convertidos". Lima-Perú, 2005. Scribd 49450705

Antonio Cruz. "Darwin no mató a Dios". Ed. Vida, Miami, Florida, 2004

Arthur C. Custance, M.A., Ph.D., "Los Restos Fósiles Del Hombre Primitivo, Y El Registro Histórico Del Génesis". Ottawa, 1968 / Rev. 1975, Artículo 45

B.A.Paramadvaiti Svami, "La Secreta Identidad De Charles Darwin", Instituto Bhaktivedanta, Srila Prabhupada. Scribd 56660937

Charles Darwin, 'Autobiografia'. Alianza Cien. pp. 85-87. Madrid, 1993.

Charles Darwin. "Autobiografía". The Project Gutenberg Etext. www.Librodot.com. Scribd 97111676

Christopher K. Mathews, K.E. van Holde & Kevin G. Ahern, "Bioquímica. Tercera Edición", Ed. Pearson Educación, Madrid, 2002

D. James Kennedy. "Por qué creo". Editorial Vida, Miami, Fl., 1982.

Daniel Durán, "¿El Hombre Primitivo?. Un desafío a la ciencia y a la teoría de la Evolución". Marzo de 2011. Scribd 50417255

Dawlin A. Ureña, Lic., "La Ciencia y la Biblia", Ed. ADF Books, Michigan, 1999

Demian Noé Cáceres Alge, etal., "Ciencia vs Religión. Una relación compleja". Scribd 43359279/200800299

Dinesh D'Souza. "Lo Grandioso del Cristianismo". Tyndale House

Publishers, Inc., Illinois, 2009.

Duane T. Gish, et al, "Creación, Evolución y Registro Fósil", Ed. Clie, Barcelona, España, 1979

Enrique Díaz Araujo, Dr, "EVOLUCIÓN Y FRAUDE", Mikael N° 7 Revista del Seminario de Paraná, Primer cuatrimestre de 1975. Scribd 25726062.

Eric J. Lerner, "The Big Bang Never Happened", Vintage Books, A Division of Random House, Inc. New York, 1991. Scribd 36326103

Eric V. Snow, "Darwin's God: Evolution and the Problem of Evil". Freetoshare Publications, 2011   Scribd 60174469

E.T. Bell. "Biografías de Grandes Matemáticos". Patricio Barros. http://www.geocities.com/grandesmatematicos/cap29.html (12 de 13) [30/10/2002 6:33:11]

"Evolución O Creacionismo ¿Quién Dice La Verdad?" Tomado De: Artículo ¿Cuán Antigua Es La Tierra? Autor: Lic. Dawlin A. Ureña; Libro Auxiliar Bíblico Portavoz. Harold L. Willmington Libro El Engaño Del Evolucionismo: Harun Yahya; Scribd 23396796

Exposición filosófica sobre el ateísmo de Nietzsche y Karl Marx. CCEH, Círculo Colimote de Estudios Hispanoamericanos. Studium I / Año I No. I Enero-Abril 1984

Frank Zorrilla, "Conociendo a Dios a través de la Ciencia", Ed. Palibrio, Bloomington, Indiana, 2011

Flory Chaves Q., "El ser como idea y la existencia de Dios en el pensamiento de Michele Federico Sciacca". Rev. Filosofía Univ. Costa Rica, XXXVII (92), 193-202, 1999

H.M. Morris, Ph.D, "Geología. ¿Actualismo o Diluvialismo?". Ed. Clie, Barcelona, España, 1980

Harold S. Slusher & Robert L. Whitelaw, "Las Dataciones Radiométricas: Crítica". Ed. Clie, Barcelona, España, 1980

HOWARD PETH, "Fe Ciega: La Evolución Expuesta" Colegio Adventista Libertad "COAL", Bucaramanga, 1996. "BLIND FAITH: Evolution Exposed", AMAZING FACTS, Inc.

P.O. Box 680, Frederick, MD 21701, USA

Henry F. Schaefer, Dr., "Los científicos y sus dioses". copyright © 1995-2003 Leadership U. Traducción de Darío Fox. © Mente Abierta 2003

Hugh Ross, "El Creador y el Cosmos", Editorial Mundo Hispano, Texas, 1999

Idara Ishaat-E-Diniyat (P) Ltd, "El Colapso De La Teoría De La Evolución En 20 Preguntas" 168/2 Jha House, Hazrat Nizamuddin Nueva Delhi - 110 013 India

Ignacio Martínez Mendizábal & Juan Luis Arsuaga Ferreras. "Amalur. Del átomo a la mente". Ediciones Temas de Hoy, S.A. (T.H.), Madrid. 2002

Indalecio Gil Albalat, "A dónde va la tierra?, Editorial Clie, Barcelona, España, 1990

Jaime Descailleaux, et al. El ADN La Molécula De La Vida. Electronic Journal Nanociencia et Moletrónica Octubre 2004, Vol. 2; N°2, (2004).

James D. Bales, Dr., "The God-killer?". Christian Crusade Publications, Tulsa, Oklahoma, 1967

José Montesinos y Sergio Toledo. "Ciencia y Religión en la Edad Moderna". Symposium «Ciencia y religión de Descartes a la Revolución Francesa» 14, 15 y 16 Septiembe 2006. Santa Cruz de la Palma
Fundación Canaria Orotava de Historia de la Ciencia, La Orotava , España

Josué Ferrer. "Por qué dejé de ser ateo". Ed. Dinámica, Florida, USA, 2009

Józef Zycinski, Mons., Arzobispo de Lublin. "Diálogo entre ciencia y fe ante las cuestiones filosóficas de la física actual" Gran Canciller de la Universidad de Lublin, Polonia. Conferencia pronunciada en un Encuentro sobre Fe y Cultura, Sevilla, 14 de marzo de 1998. http://www.unav.es/cryf/dialogoentrecienciayfe.html

Julio A. Rodríguez, IQ, "El Paradigma, ¿o cuento?, de la Evolución". Ed. Nueva Vida, New York, 2008

Kenneth Boa. "The Evolution Revolution: Naturalism and the Question of Origins". http://bible.org/seriespage/evolution-revolution-naturalism-and- question-origins

Manuel García Doncel, "El Diálogo Teología-Ciencias Hoy. Ii. Perspectivas Científica Y Teológica". Institut De Teologia Fonamental – Sant Cugat Del Vallés, enero 2003. Scribd 51771089

Mario Seiglie. "Los diez errores de Darwin". Revista "Las Buenas Noticias", Vol. 15, #3. Cincinnati, Ohio, USA. Mayo-junio, 2010

Mathews, C. K.; Van Holde, K. E.; Ahern, K. G., "BIOQUÍMICA", Pearson Educación, S.A., Madrid, 2002

Matt Ridley. "Genoma. La autobiografía de una especie". Ed. Taurus, Madrid, 2001

Michele F. Sciacca, "Mi itinerario a Cristo". Ed. Taurus (1957) Milan

Machovec. "Jesús para Ateos". Ediciones Sígueme, Salamanca, España, 1974.

Orlando Fedeli, et al. Stat Veritas. La verdad permanece. "Evolucionismo: ¿Dogma científico o Tesis teosófica?". S.Paulo, Sept. De 2003

Paul Nelson et al., "Tres puntos de vista sobre la Creación y la Evolución". Editorial Vida, Miami, FL, 2009

Paulo Arieu. "La problemática de la evolución del hombre". www.lasteologias.wordpress.com Scribd 50417255

Rafael Andrés Alemán Berenguer. "Kelvin Versus Darwin: Choque De Paradigmas En La Ciencia Decimonónica". Iluil, Vol. 33 (#71) *1er Semestre 2010 - ISSN: 0210-8615, pp. 11-24*

Rafael Llano Cifuentes. "En busca del sentido de la vida". Scribd 92740272

Saloff Astakhoff, "Origen y destino del Planeta Tierra", Editorial Clie, Barcelona, España, 1976

Samuel Vila, "A Dios por el Átomo", Ed. Clie, Barcelona, España, 1987

Scott Freeman & Jon C. Herron, "Análisis Evolutivo. Segunda Edición",
Ed. Pearson Educación, SA, Madrid, 2001

Scott M. Huse, "El Colapso de la Evolución", Ed. Chick Publications, California, 1996

Starr, Cecie & Taggart, Ralph, "BIOLOGÍA, La unidad y diversidad de la vida, décima edición", Ed. Thomson Learning, Inc., México, 2004

Stephen Jay Gould. "Desde Darwin. Reflexiones sobre historia natural".
Hermann Blume Ediciones, Madrid, 1983

Steve Keohane, "The Case for Creationism".
http://www.bibleprobe.com/creationism.htm

Ten Reasons Evolution is Wrong. Revised 3/2006.
http://www.evanwiggs.com/articles/reasons.html

Vance Ferrell, "The Evolution Handbook", Ed. Evolution Facts, Altamont,
TN, 2001

W.R. Daros, Conicet-Argentina. "El aprendizaje en la concepción, de M.F. Sciacca".

Willem Ouweneel, Ph.D, et al, "Biología y Orígenes", Ed. Clie, Barcelona, España, 1977

Additional Books written by the Author

The Missing Link - In Theology

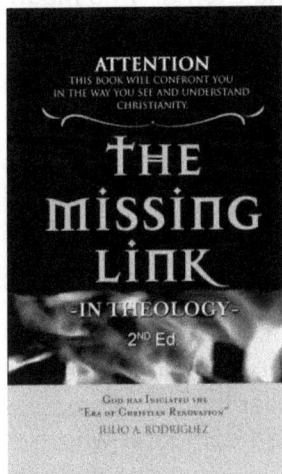

Have you ever questioned:

- What is happening to Christianity?
- Where is the glory that shone for many centuries, illuminating the minds of the souls of men?
- Why does it seem the teachings of the Bible are powerless in this generation?
- Why are there so many denominations and religions?
- Can a person be saved after they die?

You will discover non-traditional answers to these questions in the pages of this book

The Paradigm, or Tale? of Evolution

The author is a former atheist who believed in and adamantly defended the Theory of Evolution for 14 years during and after his studies at the Catholic University Pontificate Madre y Maestra in Santiago, D.R.

He graduated as a Chemical Engineer in 1978; and as a result of 30 years of

intense investigations on evolutionary theory, and innumerable life experiences he shares his conclusions.

This book details "evolutionary" thought with such impact to make you seriously question:

- Was all that exists; made by "Someone" or "Nothing"?
- Was there a Wise Being, Powerful and Eternal, who made all things; or did all begin out of "Nothingness" which never existed and never had power, a purpose...nothing at all!, formed the complete universe with its essence of nothing?
- Life? Is there a purpose or sense, or a vain illusion we must all tolerate?

The author asserts and demonstrates:

" Our schools and universities indoctrinate students against a belief in God, teaching as scientific fact purely atheistic-religiosity", and also: "If someone were to believe the universe was formed by a particle smaller than an atom, this person would have more faith than someone who professes faith in God"

Religious Gladiators. Beware of Contemporary Judaizers

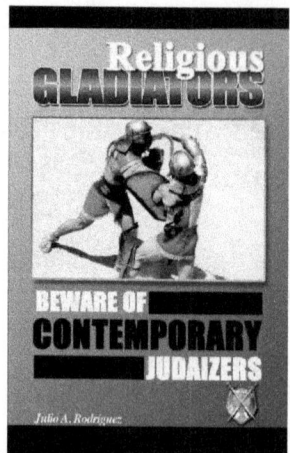

The main theme of this book is of extreme importance for Christian believers today.
There is a subtle attempt by false teachers to enter churches and lead astray the faithful, under the guise of teaching Jewish culture.

In many instances they have confused brethren lacking a solid foundation and as a consequence some have **unwittingly rejected the grace of God** manifested thru Jesus Christ; seeking justification thru the works of the law.

Other Related Information, written by the author

"Evolution is a Tall Tale and the "Big Bang" is a dogma of "faith." *A Bible vs. Science confrontation*

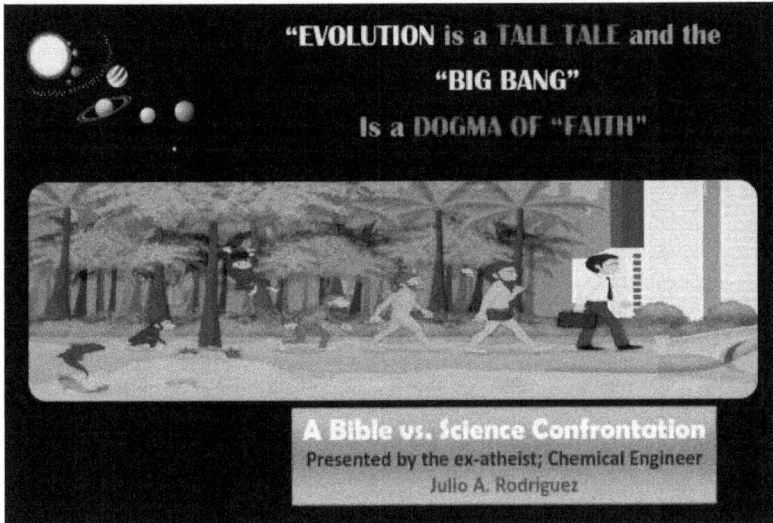

In this short presentation you will DISCOVER WHY our best students become atheists; the subtle indoctrination they receive; what blocks their use of reason; and how they are unable to think and use wisdom in analyzing all that exists.

You will be able to discern the DOGMAS OF FAITH taught in schools; convincing them to believe that it's "science". If you think you have all the answers, then respond to this great challenge. Analyze, reason, and verify the motives for your belief system.

You can find the details of this conference, in PDF,
In: http://www.scribd.com, document: 114954157

REFERENCES:

[i] Important note: Some of articles used in this compilation have been partly transcribed [shown by the symbol (...)]. If you desire to read any of the reference documents in its totality, the following information will help to locate the desired document.

[ii] http://wol.jw.org/es/wol/d/r4/lp-s/1102009552

[iii] http://www.planetseed.com/es/print/15740

[iv] Federico Sciacca, *Mi itinerario a Cristo,* Ed. Taurus, Madrid, 1957 pp. 106-115

[v] The God-Killer? Págs.120-132; 1967 © Christian Crusade Publications, Tulsa, OK

[vi] *Matt Slick* http://www.miapic.com/el-fracaso-del-ate%C3%ADsmo-para-explicar-la-moralidad%20%20%20%20

[vii] ¿Puede el hombre vivir sin Dios? Por Ravi Zacarias ©Editorial Caribe (Consecuencias del ateísmo)
http://menteabierta.es/html/articulos/ar_ysidios.htm

viii http://www.fluvium.org/textos/lectura/lectura199.htm

[ix] http://www.reasonablefaith.org/spanish/es-el-ateismo-una-filosofia-sin-esperanza

[x] *Matt Slick* http://www.miapic.com/el-fracaso-del-ate%C3%ADsmo-para-explicar-la-existencia

[xi] Frank Zorrilla, "Conociendo a Dios a través de la Ciencia", pags. 104-105, Ed. Palibrio, Bloomington, Indiana, 2011

[xii] Stephen R. Covey, "Los 7 hábitos de la gente altamente efectiva", Pags. 15-19, Ed. Paidós, Buenos Aires-Barcelona-México, 2003

[xiii] http://www.noticiacristiana.com/sociedad/2009/05/antony-flew-ateo-mas-ferreo-e-influyente-del-mundo-acepta-la-existencia-de-dios.html

[xiv] http://parameditar.wordpress.com/2009/12/21/frases-celebres-de-newton-sobre-la-existencia-de-dios/

[xv] *Daniel Iglesias Grèzes.*
http://www.feyrazon.org/PocasPal/DanError1.html

xvi http://www.explorefaith.org/speaking_collins.html

xvii http://laverdadnoshacelibres.wordpress.com/2011/07/31/francis-collins-%C2%BF-por-que-creo-en-dios/

xviii DNA and the Origin of Life: Information, Specification, and Explanation. Scribd 51569140

xix Ruby Villarreal. Juan de O'Donoju 470, Virreyes, 78240 San Luis Potosi, SLP, MEXICO creavit@terra.com

xx From the university degree textbook: "Biology, The Unity and Diversity of Life". Tenth Edition (2004). Cecie Starr/ Ralph Taggart. Págs. 326-329. Thomson, Brooks/Cole, Belmont, CA, USA.

xxi Ben Harder. Water for the Rock. Did Earth's oceans come from the heavens?. From Science News, Volume 161, No. 12, March 23, 2002. http://www.phschool.com/science/science_news/articles/water_for_the_rock.html

xxii http://www.scientificamerican.com/article.cfm?id=asteroid-killed-dinosaurs

xxiii http://news.discovery.com/animals/dinosaur-farts-120507.html

xxiv http://www.zmescience.com/science/geology/gold-meteorites-08092011/

xxv Yaiza Martínez. Tendencias Científicas. http://www.tendencias21.net/Nuestro-Universo-pudo-haberse-formado-de-otro-Universo-especular-anterior_a2195.html

xxvi http://www.euskalnet.net/ceufo/agujeros.htm

xxvii http://definicion.de/ciencia/#ixzz2CdLQ3w5r

xxviii http://www.escepticospr.com/Archivos/conocimiento_cientifico.htm

xxix http://recursostic.educacion.es/bachillerato/proyectofilosofia/version/v1/A3-1f.htm

xxx Matt Slick. http://carm.org/cut-atheism

www.ingramcontent.com/pod-product-compliance
Lightning Source LLC
Chambersburg PA
CBHW061737020426
42331CB00006B/1273